Springer INdAM Series

Volume 38

Springer INdAM Series

This series will publish textbooks, multi-authors books, thesis and monographs in English language resulting from workshops, conferences, courses, schools, seminars, doctoral thesis, and research activities carried out at INDAM - Istituto Nazionale di Alta Matematica, http://www.altamatematica.it/en. The books in the series will discuss recent results and analyze new trends in mathematics and its applications.

THE SERIES IS INDEXED IN SCOPUS

More information about this series at http://www.springer.com/series/10283

Piermarco Cannarsa • Daniela Mansutti •
Antonello Provenzale

Editors

Mathematical Approach to Climate Change and its Impacts

MAC2I

 Springer

Editors
Piermarco Cannarsa
Department of Mathematics
University of Rome "Tor Vergata"
Rome, Italy

Daniela Mansutti
Istituto per le Applicazioni del Calcolo
'M. Picone', CNR, Rome, Italy

Antonello Provenzale
Institute of Geosciences & Earth
Resources (IGG)
National Research Council (CNR)
Pisa, Italy

ISSN 2281-518X   ISSN 2281-5198 (electronic)
Springer INdAM Series
ISBN 978-3-030-38671-9   ISBN 978-3-030-38669-6 (eBook)
https://doi.org/10.1007/978-3-030-38669-6

This Springer imprint is published by the registered company Springer Nature Switzerland AG.
The registered company address is: Gewerbestrasse 11, 6330 Cham, Switzerland

In Mathematics the art of proposing a question must be held of higher value than solving it.
    (Georg Cantor)

Mathematics allows for no hypocrisy and no vagueness.
    (Stendhal)

Preface

Climate change and the associated critical problems, which might impact on the well-being and potentially the survival of humans on planet Earth, are attracting the attention of both the scientific community and the general public, sometimes with conflictual approaches. We believe that Mathematics can provide tools to address these problems, raising well-posed questions and framing them in conceptual models. The model solutions can then be analyzed and compared with observations, to provide relevant information for both diagnostic and prognostic purposes.

In this approach, it is important to fully acknowledge the difficulties associated with trying to address phenomena that are part of an extremely complex system and may still be beyond our current conceptual and computational capabilities. Such complexity, however, should not prevent us from exploring the subject using mathematical modelling, which is especially suitable to tackle issues that are still poorly understood.

The current volume was conceived in this spirit. The aim here is to provide ideas, suggestions, and motivations, inviting researchers in Applied Mathematics and related disciplines to engage and enter into dialogue with each other and with environmental scientists in this fundamental area of research.

The volume includes contributions relating to the outcomes of the international INDAM Workshop "Mathematical Approach to Climate Change Impacts—MAC2I", held in Rome at the INDAM Headquarters on 13–17 March 2017. It comprises a collection of original, peer-reviewed articles on research issues that emerged from discussions in each subject area.

The contents of this book include the following topics:

- accurate parameter determination supported by rigorous existence, regularity, uniqueness, and stability results in the case of Energy Balance Models with Memory, which are relevant to climate dynamics investigations;
- three innovative mathematical modelling approaches to complex ice state (sea ice) and composition (morainic ice), two of them related to real cases, and experimental data;

- a novel method for event screening categorization of carbon dioxide concentration time series;
- different complementary mathematical formulations of the problem of controlling the spread of invasive species, with numerical testing on real situations and data;
- influence of bedload transport processes in a coastal sand-bed river upon beach changes, with a focus on the Fiumi Uniti outlet into the Adriatic sea;
- a tutorial on mathematical modelling of large-scale motions of atmosphere and oceans, including shallowness, Earth rotation, and wind effects.

Most contributions in the book relate to actual case studies: mathematical models and/or methods are embedded into a real physical set-up, using observational data for model validation and/or addressing specific diagnostic/prognostic queries. This aspect justifies structuring the table of contents into thematic parts, namely Geophysical Fluids and Climate, Hydrology, Glaciology, and Ecosystems.

We hope that this book can be used for doctoral programs in Applied Mathematics, with courses ranging from mathematical problem-solving tools to the study of the complexities of climate and environmental change, thereby attracting critical and valuable attention to this crucial field.

Rome, Italy   Piermarco Cannarsa
Rome, Italy   Daniela Mansutti
Pisa, Italy   Antonello Provenzale
September 2019

Acknowledgements

The workshop Mathematical Approach to Climate Change Impacts—MAC2I stems from of the homonym thematic section of the excellence project of applied mathematics, MATH-TECH, funded by the Italian Ministry of Education, University and Research (MIUR) and awarded to the Istituto per le Applicazioni del Calcolo (IAC) '*M. Picone*' of the National Research Council (CNR) and to the INDAM (2015–2017) (coordinator: Roberto Natalini, IAC-CNR). Editors acknowledge the financial support for that event, of which the present scientific paper collection is a direct consequence.

Contents

Editors and Contributors

About the Editors

Piermarco Cannarsa graduated from the University of Pisa and Scuola Normale Superiore in 1979. He has been Professor of Mathematical Analysis at the University of Rome Tor Vergata since 1990. His scientific interests include partial differential equations and control theory. He has published about 130 articles in international journals as well as two books. He is President of the Italian Mathematical Union and the Italian Coordinator of the Associated International Laboratory (LIA) on Control, Optimization, Partial Differential Equations, and Scientific Computing (COPDESC).

Daniela Mansutti is Research Director at IAC-CNR. She graduated in Applied Mathematics at Sapienza University of Rome in 1984. Dr. Mansutti won a CNR fellowship for a 2-year visit at Pittsburgh University (USA) and specialized in nonlinear modelling of continuum media with K.R. Rajagopal. She then became a research scientist at IAC. She has authored about 60 papers in peer-reviewed international ISI journals. As coordinator or WP principal investigator on national/transnational research calls, she has studied problems in several physical contexts (newtonian/non-newtonian flows, phase change, glaciology, ecosystems etc.) with partial differential equations numerical modelling and stability analysis.

Antonello Provenzale is the Director of IGG-CNR. His research areas are geosphere-biosphere interaction, climate dynamics, and ecosystems. He is a recipient of the Golden Badge Award of the European Geophysical Society. Dr. Provenzale has been an invited professor at ENS and Pierre and Marie Curie University in Paris, Ben Gurion University in Israel, and University of Colorado in Boulder (USA). He is Coordinator of the EU H2020 project "ECOPOTENTIAL" and WP Lead in the H2020 project "eLTER Plus", as well as Coordinator of the Global Ecosystem Initiative of GEO. He is the author of more than 150 papers in peer-reviewed international ISI journals.

Contributors

Stefano Bianchi Department of Mathematics and Physics, Roma Tre University, Rome, Italy

Paolo Billi IPDRE, Tottori University, Tottori, Japan

Enrico Calzavarini Université de Lille, Unité de Mécanique de Lille, UML EA 7512, Lille, France

Piermarco Cannarsa Dipartimento di Matematica, Università di Roma "Tor Vergata", Rome, Italy

Francesca Casella Istituto di Scienze delle Produzioni Alimentari, CNR, Bari, Italy

Paolo Ciavola Department of Physics and Earth Sciences, University of Ferrara, Ferrara, Italy

Silvia Cilli Department of Physics and Earth Sciences, University of Ferrara, Ferrara, Italy

Fasma Diele Istituto per le Applicazioni del Calcolo 'M. Picone', CNR, Bari, Italy

Alcide Giorgio di Sarra Laboratory for Observations and Analyses of the Earth and Climate, ENEA, Rome, Italy

Michael Ghil Ecole Normale Superieure, Paris, France
University of California, Los Angeles, CA, USA

Edoardo Grottoli Department of Physics and Earth Sciences, University of Ferrara, Ferrara, Italy

Krishna Kannan Department of Mechanical Engineering, Indian Institute of Technology, Madras, Chennai, India

Deborah Lacitignola Dipartimento di Ingegneria Elettrica e dell'Informazione, Università di Cassino e del Lazio meridionale, Cassino, Italy

Reza Malek-Madani Department of Mathematics, US Naval Academy, Annapolis, MD, USA

Martina Malfitana Dipartimento di Matematica, Università di Roma "Tor Vergata", Roma, Italy

Daniela Mansutti Istituto per le Applicazioni del Calcolo 'M. Picone', CNR, Rome, Italy

Carmela Marangi Istituto per le Applicazioni del Calcolo 'M. Picone', CNR, Bari, Italy

Patrick Martinez Institut de Mathématiques de Toulouse, UMR 5219, Université de Toulouse, CNRS UPS IMT, Toulouse Cedex, France

Angela Martiradonna Istituto per le Applicazioni del Calcolo 'M. Picone', CNR, Bari, Italy

Salvatore Piacentino Laboratory for Observations and Analyses of the Earth and Climate, ENEA, Rome, Italy

Wolfango Plastino Department of Mathematics and Physics, Roma Tre University, Rome, Italy

Antonello Provenzale Istituto di Geoscienze e Georisorse, CNR, Pisa, Italy

Stefania Ragni Department of Economics and Management, University of Ferrara, Ferrara, Italy

Kumbakonam R. Rajagopal Department of Mechanical Engineering, Texas A&M University, College Station, TX, USA

Andrea Scagliarini Istituto per le Applicazioni del Calcolo 'M. Picone', CNR, Rome, Italy

Leonardo Schippa Department of Engineering, University of Ferrara, Ferrara, Italy

Damiano Sferlazzo Laboratory for Observations and Analyses of the Earth and Climate, ENEA, Rome, Italy

Eric Simonnet Institut de Physique de Nice, Universite Cote d'Azur, Sophia Antipolis, France

Federico Toschi Eindhoven University of Technology, Eindhoven, The Netherlands
Istituto per le Applicazioni del Calcolo 'M Picone', CNR, Rome, Italy

Stefano Urbini Istituto Nazionale di Geofisica e Vulcanologia, Rome, Italy

Part I
Theme: Geophysical Fluids and Climate

Geophysical Fluid Dynamics, Nonautonomous Dynamical Systems, and the Climate Sciences

Michael Ghil and Eric Simonnet

Abstract This contribution introduces the dynamics of shallow and rotating flows that characterizes large-scale motions of the atmosphere and oceans. It then focuses on an important aspect of climate dynamics on interannual and interdecadal scales, namely the wind-driven ocean circulation. Studying the variability of this circulation and slow changes therein is treated as an application of the theory of nonautonomous dynamical systems. The contribution concludes by discussing the relevance of these mathematical concepts and methods for the highly topical issues of climate change and climate sensitivity.

Keywords Bifurcations · Climate sensitivity · Double-gyre problem · Low-frequency variability · Pullback and random attractors · Wasserstein distance · Wind-driven ocean circulation

1 Effects of Rotation

The first two chapters of this contribution are dedicated to an introductory review of the effects of rotation and shallowness on large-scale planetary flows. The theory of such flows is commonly designated as geophysical fluid dynamics (GFD), and it applies to both atmospheric and oceanic flows, on Earth as well as on other planets. GFD is now covered, at various levels and to various extents, by several books [35, 56, 70, 105, 120, 134, 164].

M. Ghil (✉)
Ecole Normale Supérieure and PSL University, Paris, France

University of California, Los Angeles, CA, USA
e-mail: ghil@lmd.ens.fr

E. Simonnet
Institut de Physique de Nice, CNRS & Université Côte d'Azur, Nice, Sophia-Antipolis, France
e-mail: eric.simonnet@inphyni.cnrs.fr

© Springer Nature Switzerland AG 2020
P. Cannarsa et al. (eds.), *Mathematical Approach to Climate Change and Its Impacts*, Springer INdAM Series 38,
https://doi.org/10.1007/978-3-030-38669-6_1

The virtue, if any, of this presentation is its brevity and, hopefully, clarity. It follows most closely, and updates, Chapters 1 and 2 in [56]. The intended audience includes the increasing number of mathematicians, physicists and statisticians that are becoming interested in the climate sciences, as well as climate scientists from less traditional areas—such as ecology, glaciology, hydrology, and remote sensing—who wish to acquaint themselves with the large-scale dynamics of the atmosphere and oceans.

For tutorial purposes, we start this chapter with the dynamical effects of rotation on large-scale flows in the simplest cases. The Rossby number is defined as a nondimensional measure of the importance of rotation, and the usefulness of a rotating frame of reference is outlined.

Equations of fluid motion are first stated and discussed in an inertial frame of reference. The necessary modifications for passing to a rotating frame are derived. The Coriolis acceleration is shown to be the main term that distinguishes between the equations in a rotating and an inertial frame.

Vorticity is defined and approximate conservation laws for it, namely Kelvin's theorem and Ertel's theorem, are derived. The important distinction between barotropic and baroclinic flows is made. Finally, motion at small Rossby number is discussed, the geostrophic approximation is stated, and the Taylor-Proudman theorem for barotropic flow is derived from it. Geostrophic flow is shown to be a good qualitative approximation for atmospheric cyclones.

1.1   The Rossby Number

Newtonian mechanics recognizes a special class of coordinate systems, called inertial frames, and by rotation we always mean rotation relative to an inertial frame. As we shall see, Newtonian fluid dynamics relative to a rotating frame have a number of unusual features.

For example, an observer mowing with constant velocity in an inertial frame feels no force. If the observer, however, has a constant velocity with respect to a frame attached to a rotating body, a Coriolis force will generally be felt, acting in a direction perpendicular to the velocity. Such "apparent" forces lead to dynamical results which are unexpected and peculiar when compared with conventional dynamics in an inertial frame. We should therefore begin by verifying that this shift of viewpoint to a noninertial frame is really called for by the physical phenomena to be studied, as well as by our individual, direct perception of these flows, as rotating observers.

To examine this question for the dynamics of fluids on the surface of the Earth, we define Ω to be the Earth's angular speed, so $2\pi/\Omega \cong 24\,$h. Therefore, at a radius of $6380\,$km a point fixed to the equator and rotating with the surface moves with speed $1670\,$km/h as seen by an (inertial) observer at the center of the Earth. On the other hand, the large-scale motion of the fluid relative to the solide surface never exceeds a speed of about $300\,$km/h, which occurs in the atmospheric jet streams. Thus the speeds relative to the surface are but a fraction of inertial speeds, and it

Table 1 Rossby number values r for the Earth and for Jupiter, and for different characteristic length scales

r	Feature	L	U	ε
Earth	Gulf Stream	100 km	1 m/s	0.07
$\Omega = 7.3 \cdot 10^{-5} \mathrm{s}^{-1}$	Weather system	1000 km	20 m/s	0.14
	Core	3000 km	0.1 cm/s	$2 \cdot 10^{-7}$
Jupiter				
$\Omega = 1.7 \cdot 10^{-4} \mathrm{s}^{-1}$	Bands, Red Spot	10^4 km	50 m/s	0.015

will be useful to remove the fixed component of motion due to the Earth's rotation, by introducing equations of motion relative to rotating axis.

In this rotating frame, we define U to be a characteristic speed relative to the surface, and define L as a length over which variations of relative speed of order U occur. From these parameters we then obtain a characteristic time $T = L/U$, which is a time scale for the evolution of the fluid structure in question. Since Ω is an inverse time, the product $T\Omega$, or more customarily the *Rossby number*

$$\varepsilon = \frac{U}{2\Omega L} \tag{1}$$

is a dimensionless measure of the importance of rotation, when assessed over the lifetime or period of the structure.

The nature of this rotational effect will be given a dynamical basis forthwith. For the moment, it is important only to realize that small ε means that the effects of rotation are important. With this in mind, we show in Table 1 values of the Rossby number for a few large-scale flows on the Earth and Jupiter.

We conclude that rotation is important in all of these systems and that a rotating coordinate system is probably useful for their study. Minor changes in these estimates are needed to account for the fact that the proper measure of rotation is the local angular velocity of the tangent plane to the planet, as shown in Sect. 1.2 below.

Also, these estimates only mean that Coriolis forces, proportional to $2\Omega U$, are typically larger than the forces needed to produce the acceleration measured relative to the rotating frame, these being of magnitude U^2/L (or U/T) times the fluid density. There may be other forces, such as viscous or magnetic forces, which enter into the dynamical balance. In the Earth's fluid core, for example, magnetic and Coriolis forces are probably comparable, while in the free atmosphere and ocean it is predominantly the "pressure" forces which compete with the Coriolis force in the dominant dynamical balance.

1.2 Equations of Motion in an Inertial Frame

We consider a fluid of density ρ, moving under the influence of a pressure field $p(r, t)$ and a body force $F(r, t)$, where $r = (x, y, z) = (x_1, x_2, x_3)$ is the position vector and t is time. The equations describing the *local conservation of momentum and mass* are

$$\frac{\partial \mathbf{u}}{\partial t} + (\mathbf{u} \cdot \nabla)\mathbf{u} = -\frac{1}{\rho}\nabla p + \frac{1}{\rho}\mathcal{F}, \tag{2a}$$

$$\frac{\partial \rho}{\partial t} + \nabla \cdot \rho\mathbf{u} = 0, \tag{2b}$$

where $\mathbf{u}(r, t)$ is the velocity field and $\nabla = (\partial_x, \partial_y, \partial_z)$ is the gradient operator, with bold type used for vectors. Equation (2b) is often called either the "continuity equation" or the "mass conservation equation."

For a given $\mathcal{F}$, (2) comprises four scalar equations for the three velocity components of $\mathbf{u}$, the density ρ and the pressure p. Another relation is therefore needed to close the system, e.g., an equation of state connecting p and ρ.

In many problems of geophysical interest, such as ocean dynamics, the approximate incompressibility of the fluid plays a prominent role. In these cases, $\mathbf{u}$ is prescribed to be a solenoidal, i.e., divergence-free, vector field, and thus it closes the system. For the moment, we leave the closure unspecified and deal with the incomplete system (2). We also take $\mathcal{F}$ at first to be a conservative field, such as gravitation, defined in terms of a potential ϕ by

$$\mathcal{F} = -\rho\nabla\phi. \tag{3}$$

We are thus excluding here the possibility of the important contribution of the nonconservative viscous stresses.

The operator

$$\frac{D}{Dt} \equiv \frac{\partial}{\partial t} + \mathbf{u} \cdot \nabla \tag{4}$$

occurring in (2) is called the *material derivative* or the *Lagrangian derivative*, since it is the time derivative calculated by an observer moving with the local or material velocity. In the following, we will use the notation D/Dt rather than d/dt when appropriate to avoid any confusion. To verify this last claim let $G(\mathbf{r}, t)$ be a field and let $\mathbf{r} = \mathbf{R}(t)$ be the position of a fluid particle at time t. Then $\mathbf{u} = (u, v, w) = (u_1, u_2, u_3) = \frac{d\mathbf{R}}{dt}$ is the velocity of the particle, and we have

$$\frac{d}{dt}G\left(\mathbf{R}\left(t\right), t\right) = \frac{\partial G}{\partial t} + \frac{d\mathbf{R}}{dt} \cdot \nabla G = \frac{\partial G}{\partial t} + u_i\frac{\partial G}{\partial x_i}$$

where the summation convention of repeated indices is used. A quantity Q (scalar, vector or tensor) which satisfies $DQ/Dt = 0$ is said to be a *material invariant* because the quantity $Q(\mathbf{R}(t), t)$ just stays constant along the trajectory defined by $\mathbf{R}$.

1.2.1 Thermodynamics

Very often, the equation of state does not involve the pressure and density alone. Typically, one often has

$$\rho_{ocean} = \rho(S, T, p), \quad \rho_{atmos} = \rho(T, p), \tag{5}$$

where S is the salinity and T the temperature. One needs, therefore, two additional equations for T (and S). In the following, we consider the oceanic case, although the cases of both a dry and a humid atmosphere can be treated in an analogous way.

We will suppose later on that the fluid is in a local thermodynamic equilibrium. This assumption implies that the variables entering Eq. (6) below are locally related or that they can be simply expressed in mathematical form:

$$E = E(\alpha, \eta, S), \tag{6}$$

where E is the *specific internal energy*, $\alpha \equiv 1/\rho$ is called the *specific volume*, η the *specific entropy* and S the salinity. Taking the total differential of (6),

$$dE = \frac{\partial E}{\partial \alpha} d\alpha + \frac{\partial E}{\partial \eta} d\eta + \frac{\partial E}{\partial S} dS, \tag{7}$$

enables one to interpret it in a illuminating way: The quantity $-\partial E(\alpha, \eta, S)/\partial \alpha$ is interpreted as the pressure p, while the term $(\partial E/\partial \alpha)d\alpha$ is the change in energy due to mechanical compression. Note that $d\alpha < 0$ and $p > 0$ give $dE > 0$, i.e. energy must be supplied to reduce volume if fluid resists compression.

Similarly, $\partial E(\alpha, \eta, S)/\partial \eta$ is interpreted as the temperature T, and the term $(\partial E/\partial \eta)d\eta$ is the energy change due to heating. Finally, $\partial E(\alpha, \eta, S)/\partial S$ is called the *chemical potential* μ of seawater. One has therefore

$$dE = -pd\left(\frac{1}{\rho}\right) + Td\eta + \mu dS. \tag{8}$$

It appears that the salinity and specific entropy are conserved along particle motion when there are no heat sources and no molecular diffusion, i.e.

$$\frac{D\eta}{Dt} = \frac{DS}{Dt} = 0. \tag{9}$$

In order to have an equation involving T instead of η, one expresses the entropy η as a function of p, S and T only. It is possible to eliminate α between $p(\alpha, \eta, S)$ and $T(\alpha, \eta, S)$. Using (9), one has

$$\frac{D\eta}{Dt} = \frac{\partial \eta}{\partial p}\frac{Dp}{Dt} + \frac{\partial \eta}{\partial T}\frac{DT}{Dt} + \frac{\partial \eta}{\partial S}\frac{DS}{Dt} = \frac{\partial \eta}{\partial p}\frac{Dp}{Dt} + \frac{\partial \eta}{\partial T}\frac{DT}{Dt} = 0. \tag{10}$$

It appears that the partial derivatives are known quantities:

$$\left(\frac{\partial \eta}{\partial T}\right)_p \equiv \frac{C_p}{T} \quad \text{and} \quad \left(\frac{\partial \eta}{\partial p}\right)_T \equiv -\frac{C_T}{\rho}, \tag{11}$$

where C_p is the *specific heat* at constant pressure and salinity and C_T the coefficient of thermal expansion. Therefore, one ends up with

$$\frac{DT}{Dt} = \frac{T}{\rho} \frac{C_T}{C_p} \frac{Dp}{Dt}, \tag{12a}$$

$$\frac{DS}{Dt} = 0. \tag{12b}$$

For the oceans, the right-hand side of the first equation is often neglected, unless pressure variations are very large compared to temperature fluctuations, as could be the case in the deep ocean. Note that alternative formulations to close the system (2) do exist, depending on which variables one considers, instead of the temperature T addressed here. The overall procedure, however, is always the same and it makes use of the internal energy (6) and of the identities (7) and (9).

Example In the case of the atmosphere, the ideal gas law is a good approximation of the equation of state for dry air.

$$\rho = \frac{p}{RT}. \tag{13}$$

The specific entropy η is

$$\eta = C_p \log T - R \log p. \tag{14}$$

One obtains $C_T = -\rho \left(\frac{\partial \eta}{\partial p}\right)_T = \frac{1}{T}$.[1] Plugging this expression in (12a) and using (13), one obtains

$$\frac{DT}{Dt} = \frac{R}{C_p} \frac{T}{p} \frac{Dp}{Dt}, \tag{15}$$

or with straightforward algebraic manipulations

$$\frac{D\theta}{Dt} = 0, \quad \text{with } \theta = T \left(\frac{p_0}{p}\right)^{R/C_p}. \tag{16}$$

---

[1]Note that, since only the variables T and p are involved, we could simply derive (14) by integrating (11); in this case, one obtains $T C_T = \text{const}$.

The quantity θ is called the *potential temperature* and p_0 a constant pressure reference.

1.3 Equations in a Rotating Frame

We would like to use the form taken by (2) relative to a rotating coordinate system. That is, we would like to work with the velocity as perceived by an observer fixed in the rotating frame, and with time differentiation as would be performed by such an observer. Spatial differentiation, involving a limit process at fixed time, is an operation which is invariant under transformation to a rotating frame. The essential problem is thus to deal properly with time derivatives.

1.3.1 Absolute and Relative Derivative

Since differentiation involves a limit process, it is convenient to begin with an infinitesimal rotation of the axis. Consider a constant vector $\mathbf{P}$ fixed rigidly relative to the rotating frame. Let the frame rotates with respect to the inertial frame through a small angle $d\theta$ about the z or x_3-axis. If $\mathbf{P}$ has coordinates ($P_1 = a\cos\varphi$, $P_2 = a\sin\varphi$, P_3) in the rotated frame, an *inertial* observer will measure in his frame, after rotation, some coordinates ($P_1' = a\cos(\varphi + d\theta)$, $P_2' = a\sin(\varphi + d\theta)$, $P_3' = P_3$). Using the smallness of $d\theta$ one easily obtains ($P_1' = P_1 - P_2 d\theta$, $P_2' = P_2 + P_1 d\theta$, $P_3' = P_3$) for $\mathbf{P}$. Thus, under this rotation, we have the invariant vector relation

$$d\mathbf{P} = d\theta \times \mathbf{P}, \tag{17}$$

where the vector $d\theta = (0, 0, d\theta)$. Introducing time t, we write (17) as

$$\left(\frac{d\mathbf{P}}{dt}\right)_0 = \frac{d\theta}{dt} \times \mathbf{P} \equiv \Omega \times \mathbf{P}, \tag{18}$$

where we have defined $\Omega(t) = (0, 0, d\theta/dt)$ as the instantaneous angular velocity of the rotating frame. Equation (18) involves the subscript "0", which will stand for "as determined by an observer in the inertial frame", sometimes called *absolute*. Since $\mathbf{P}$ is a constant vector in the rotating frame, (18) simply states that, in this case, the inertial observer sees a vector attached to the origin rotating (about the z-axis. Examples of such "rotating, fixed vectors" are a set of orthonormal basis vectors in the rotating frame, ($\mathbf{i_1}, \mathbf{i_2}, \mathbf{i_3}$), $\mathbf{i_k}$ being the unit vector along the x_k axis. Thus, we can recompute (18) as follows. Define

$$\mathbf{P} = P_1\mathbf{i_1} + P_2\mathbf{i_2} + P_3\mathbf{i_3} \tag{19}$$

as the representation of $\mathbf{P}$ in the rotating basis. Since P_1, P_2, and P_3 are independent of time we have, using (18) and the summation convention,

$$\left(\frac{d\mathbf{P}}{dt}\right)_0 = P_k \frac{d\mathbf{i_k}}{dt} = P_k \left(\Omega \times \mathbf{i_k}\right) = \Omega \times P_k \mathbf{i_k} = \Omega \times \mathbf{P}. \tag{20}$$

Thus we see that the basis vectors carry the effect of rotation in their time dependence.

Suppose now that the coordinates P_k in (19) depend on time, $P_k = P_k(t)$. Repeating (20) we obtain

$$\left(\frac{d\mathbf{P}}{dt}\right)_0 = \left(\frac{d\mathbf{P}}{dt}\right)_r + \Omega \times \mathbf{P}, \tag{21a}$$

where

$$\left(\frac{d\mathbf{P}}{dt}\right)_r = P_k'(t)\,\mathbf{i_k} \tag{21b}$$

is the time derivative of $\mathbf{P}$ relative to the rotating observer, with $P_k' = dP_k(t)/dt$. The relation (21) is the basic result that must be used to transform the equations of motion to a rotating frame of reference.

We remark that Ω is allowed to depend on time t in (21). This can be of interest in geophysical fluid dynamics when modeling the precession of the Earth's axis of rotation However, this time dependence has negligible effects for most phenomena in the dynamics of the atmosphere, oceans, or fluid core of the earth and will be largely ignored below. We also note that, for the purpose of interpreting a formula such as (18) or (21), it is sometimes convenient to think of the relation as applying at an instant when the two frames of reference coincide.

Actually, we do not need (21) for the continuity equation (2b), since it is obvious that the material derivative of a *scalar* field is independent of the coordinate frame. Indeed, its value is determined by the time dependence observed for a given particle of fluid. To verify the invariance explicitly, let $G(\mathbf{r}, t)$ be a scalar field as observed in the inertial frame.

Then, for *partial* time derivatives we have $\mathbf{P} = \mathbf{r}$ in (18) and thus

$$\left(\frac{\partial G}{\partial t}\right)_r = \left(\frac{\partial G}{\partial t}\right)_0 + (\Omega \times \mathbf{r}) \cdot \nabla G. \tag{22a}$$

Now, applying (21) to the vector $\mathbf{r} = \mathbf{R}(t)$, which represents the particle position at time t, yields

$$\mathbf{u}_0 = \left(\frac{d\mathbf{R}}{dt}\right)_0 = \left(\frac{d\mathbf{R}}{dt}\right)_r = \mathbf{u}_r + \Omega \times \mathbf{r}, \tag{22b}$$

where

$$\mathbf{u}_r = \left(\frac{d\mathbf{R}}{dt}\right)_r \tag{22c}$$

is the velocity vector relative to the rotating frame. Using (22a), (22b) one obtains

$$\left(\frac{\partial G}{\partial t}\right)_r + \mathbf{u}_r \cdot \nabla G = \left(\frac{\partial G}{\partial t}\right)_0 + \mathbf{u}_0 \cdot \nabla G,$$

which confirms the invariance of the material derivative.

1.3.2 Relative Acceleration and Coriolis Acceleration

The right-hand side of (22b) is another vector function of t, and we may repeat the process to obtain the acceleration:

$$\left(\frac{d^2\mathbf{R}}{dt^2}\right)_0 = \frac{d}{dt}[(\mathbf{u}_r + \Omega \times \mathbf{r})]_r + \Omega \times [\mathbf{u}_r + \Omega \times \mathbf{r}]$$

$$= \left(\frac{d\mathbf{u}_r}{dt}\right)_r + 2\Omega \times \mathbf{u}_r + \dot{\Omega} \times \mathbf{r} + \Omega \times (\Omega \times \mathbf{r}) \tag{23}$$

$$= \left(\frac{d\mathbf{u}_0}{dt}\right)_0.$$

where $\mathbf{u}_0$ is the velocity perceived by the inertial observer. Using (23) and (2a) and recalling the invariance of spatial derivatives we thus obtain the desired system of equations relative to the rotating frame:

$$\left[\left(\frac{\partial \mathbf{u}_r}{\partial t}\right)_r + (\mathbf{u}_r \cdot \nabla)\mathbf{u}_r + 2\Omega \times \mathbf{u}_r + \dot{\Omega} \times \mathbf{r} + \Omega \times (\Omega \times \mathbf{r})\right] = -\frac{1}{\rho}\nabla p - \nabla \phi \tag{24a}$$

$$\left(\frac{\partial \rho}{\partial t}\right)_r + \nabla \cdot \rho \mathbf{u}_r = 0. \tag{24b}$$

If Ω is taken to be constant, i.e. $\dot{\Omega} = 0$ and if we use the identity

$$\Omega \times (\Omega \times \mathbf{r}) = -\nabla\left(\frac{|\Omega \times \mathbf{r}|^2}{2}\right), \tag{25}$$

then Eq. (24a) takes the form

$$\left[\left(\frac{\partial \mathbf{u}_r}{\partial t}\right)_r + (\mathbf{u}_r \cdot \nabla)\mathbf{u}_r + 2\Omega \times \mathbf{u}_r\right] = -\frac{1}{\rho}\nabla p - \nabla \phi_c \tag{26a}$$

$$\phi_c = \phi - \frac{1}{2}\mid \Omega \times \mathbf{r}\mid^2, \tag{26b}$$

where ϕ_c now absorbs the *centripetal acceleration*. The latter is only about 1/300 of the gravitational acceleration at the surface of the Earth and so is negligible for most atmospheric and oceanographic purposes. The remaining term which distinguishes Eq. (26) for the relative acceleration from Eq. (2a) for the absolute acceleration is the *Coriolis acceleration* $2\Omega \times \mathbf{u}_r$.

If U and L are the characteristic scales of Sect. 1.1 defined now in the rotating system, and if L/U is taken as the characteristic time, then in order of magnitude

$$\left(\frac{d\mathbf{u}_r}{dt}\right)_r \sim U^2/L, \quad 2\Omega \times \mathbf{u}_r \sim 2\Omega U, \tag{27}$$

and so the Rossby number (1) has the dynamical meaning of a characteristic ratio of the relative acceleration of a fluid element to the Coriolis acceleration.

For future reference, we now drop the subscript "r" and rewrite the basic equations (24) as:

$$\frac{\partial \mathbf{u}}{\partial t} + (\mathbf{u} \cdot \nabla)\mathbf{u} + 2\Omega \times \mathbf{u} = -\frac{1}{\rho}\nabla p - \nabla \phi_c - \dot{\Omega} \times \mathbf{r}$$

$$\frac{\partial \rho}{\partial t} + \nabla \cdot \rho \mathbf{u} \qquad\qquad = 0. \tag{28}$$

1.4 Vorticity

For any velocity field $\mathbf{u}$ we define the associated vorticity field by

$$\boldsymbol{\omega} = \nabla \times \mathbf{u}, \tag{29}$$

where we recall the explicit calculation:

$$\nabla \times \mathbf{u} = \begin{pmatrix} \partial_x \\ \partial_y \\ \partial_z \end{pmatrix} \times \begin{pmatrix} u \\ v \\ w \end{pmatrix} = \begin{pmatrix} \partial_y w - \partial_z v \\ \partial_z u - \partial_x w \\ \partial_x v - \partial_y u \end{pmatrix}. \tag{30}$$

Remark The vorticity is a divergence-free vector field: $\nabla \cdot \boldsymbol{\omega} = \nabla \cdot \nabla \times \mathbf{u} = 0$. In particular, thanks to Stoke's theorem, for any surface S enclosing a volume V in $\mathbb{R}^3$

Fig. 1 The circulations C_1 and C_2 around a vortex tube are the same. The strength of a vortex tube is constant along the length of the tube

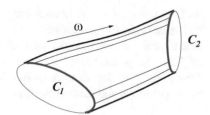

(for instance a sphere), one has

$$\int_S \boldsymbol{\omega} \cdot \mathbf{dS} = \int_V \nabla \cdot \boldsymbol{\omega} \, \mathbf{dV} = 0. \tag{31}$$

Define now a *vortex line* as the integral curve of the vorticity field $\boldsymbol{\omega}$ meaning that any points on the vortex line is tangent to the vorticity vector field. A *vortex tube* is composed of vortex lines going through a closed curve (see Fig. 1). Now applying (31) to a portion of this vortex tube determined by two different closed curves C_1 and C_2, one obtains the results that the circulation is the same around C_1 and C_2 (see Fig. 1). Indeed the lateral surface between C_1 and C_2 does not contribute to the circulation since it is tangent to the vector field. One can thus write

$$\int_{C_1} \mathbf{u} \cdot \mathbf{dr} = \int_{C_2} \mathbf{u} \cdot \mathbf{dr}. \tag{32}$$

Note that this is merely a "static" property since the vortex tube is frozen (t is fixed). A dynamical view will be provided by Kelvin's theorem. $\square$

For solid body rotation with angular velocity Ω, the velocity field is obtained by setting $\mathbf{u}_r = 0$, so that (cf. (22b)) $\mathbf{u} = \mathbf{u}_0 = \Omega \times \mathbf{r}$. The associated vorticity is

$$\nabla \times (\Omega \times \mathbf{r}) = \Omega (\nabla \cdot \mathbf{r}) - \Omega \cdot \nabla \mathbf{r} = 3\Omega - \Omega = 2\Omega. \tag{33}$$

Applied locally in an arbitrary velocity field, vorticity has the same meaning: it is equal to twice the angular velocity of a fluid element.

This interpretation of vorticity in terms of a local angular velocity is misleading in some respects, since the term "fluid element" is not really appropriate for describing what is basically a point property. More precisely, the rate-of-strain tensor given by the matrix of partial derivatives of the components of $\mathbf{u}$ at any point has a symmetric and an antisymmetric part. Vorticity is associated with the antisymmetric part, while the symmetric part gives rise to a *straining field*. The effect of the latter is to distort "fluid elements", by differentially stretching and compressing them. With this caveat, we return to the discussion of vorticity.

Since global solid body rotation can be characterized, according to (33), as a state of uniform vorticity, the mechanics of the fluid relative to a *rotating* coordinate system can alternatively be regarded as a mechanics of deviations from a state of uniform vorticity. As we have already noted, the practical usefulness of reference to a state of uniform rotation will depend on the magnitude of the deviations, and the Rossby number can also be regarded as a *measure of relative vorticity as a fraction of* 2Ω. The properties of vorticity relative to the rotating frame are thus clearly "inherited" from the mechanics in the inertial frame, so for the present discussion we return to (2) and omit subscripts.

1.4.1 Kelvin's Theorem

We now use the vector identity (please check by using formula (30))

$$(\mathbf{u} \cdot \nabla)\mathbf{u} = -\mathbf{u} \times \omega + \nabla \left(\frac{1}{2}|\mathbf{u}|^2 \right) \tag{34}$$

so that, (2a) can be reformulated as

$$\frac{\partial \mathbf{u}}{\partial t} + \nabla \left(\frac{1}{2}|\mathbf{u}|^2 \right) - \mathbf{u} \times \omega + \frac{1}{\rho}\nabla p = -\nabla\phi. \tag{35}$$

The particularly simple case in which $\mathbf{u} \times \omega = 0$ is called a Beltrami flow. Taking the curl of (35) we obtain an equation for the vorticity,

$$\frac{\partial \omega}{\partial t} + (\mathbf{u} \cdot \nabla)\omega - (\omega \cdot \nabla)\mathbf{u} + \omega\nabla \cdot \mathbf{u} - \frac{1}{\rho^2}\nabla\rho \times \nabla p = 0. \tag{36a}$$

Using (2b) to eliminate $(\nabla \cdot \mathbf{u})$ from (36a) we have

$$\frac{\partial \omega}{\partial t} + (\mathbf{u} \cdot \nabla)\omega - (\omega \cdot \nabla)\mathbf{u} - \frac{\omega}{\rho}\frac{\mathrm{D}\rho}{\mathrm{D}t} - \frac{1}{\rho^2}\nabla\rho \times \nabla p = 0. \tag{36b}$$

Multiplying (36b) by $1/\rho$ and rearranging we obtain finally a simpler equation for ω/ρ,

$$\frac{\mathrm{D}(\omega/\rho)}{\mathrm{D}t} = \frac{\omega}{\rho} \cdot \nabla\mathbf{u} + \frac{1}{\rho^3}\nabla\rho \times \nabla p. \tag{37}$$

It shows that the two terms on the right-hand-side will tend to change ω/ρ as one moves with the fluid. In particular, the vector ω/ρ is not in general a material invariant.

It would appear that the vorticity vector itself does not provide one with conservation laws having an intuitive appeal, basically because of the already mentioned non-local nature of physical reasoning based upon "fluid elements." It is advantageous, therefore, to recognize the divergence-free (solenoidal) property

Fig. 2 Kelvin's theorem: the circulation Γ_t at time t is the same than the circulation Γ_{t+dt} at time $t + dt$ provided the fluid is barotropic and the corresponding closed contours consists of material points. Note that the vortex tube is displaced by the flow but might not be a vortex tube anymore (the vortex tube at time $t + dt$ is shown in thin dashed contours)

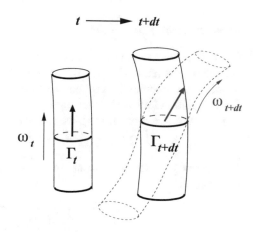

of the vorticity vector and to exploit the non-local concept of *flux* of a divergence-free vector field. This non-local approach is the basis for the fundamental theorem discovered by Lord Kelvin. *Kelvin's theorem* studies the evolution of the flux of ω through a surface S bounded by a simple closed contour C, when the latter consists of material points, i.e. of points which move with the fluid (see Fig. 2).

By Stoke's theorem, the flux we study is given by

$$\int_S \omega \cdot \mathbf{ds} = \int_C \mathbf{u} \cdot \mathbf{dr}, \tag{38a}$$

where

$$\Gamma(t) = \int_C \mathbf{u} \cdot \mathbf{dr} \tag{38b}$$

is called the *circulation* on C. Let C be given by parametric equations $\{\mathbf{r} = \mathbf{R}(\alpha, t), 0 \leq \alpha < 1\}$; then a given material point in the curve corresponds to a given value of α. The parameter α therefore is a material of Lagrangian parameter. One may think of C as a smoke ring for instance. One obtains

$$
\begin{aligned}
\frac{D\Gamma}{Dt} &= \frac{D}{Dt} \int_C \mathbf{u} \cdot \frac{\partial \mathbf{R}}{\partial \alpha} d\alpha \\
&= \int_C \frac{D\mathbf{u}}{Dt} \cdot \mathbf{dr} + \int_C \mathbf{u} \cdot \frac{D}{Dt}\left(\frac{\partial \mathbf{R}}{\partial \alpha}\right) d\alpha \\
&= \int_C \frac{D\mathbf{u}}{Dt} \cdot \mathbf{dr} + \int_C \mathbf{u} \cdot \frac{\partial \mathbf{u}}{\partial \alpha} d\alpha \\
&= \int_C [-\frac{\nabla p}{\rho} - \nabla \phi] \cdot \mathbf{dr} = -\int_C \frac{\nabla p}{\rho} \cdot \mathbf{dr}, \tag{39}
\end{aligned}
$$

where the term $\int_C \mathbf{u} \cdot \dfrac{\partial \mathbf{u}}{\partial \alpha} d\alpha = \int_C d(|\mathbf{u}|^2/2) = 0$ vanishes on a closed contour and the material derivative of $\mathbf{r} = \mathbf{R}(\alpha, t)$ is just $\mathbf{u}$, i.e. $\mathbf{u} = \dfrac{D\mathbf{r}}{Dt}$. Note that the time derivative is a material derivative here and since C is moving with the fluid it is possible to commute integration and time derivative as done in the second line of this calculation letting C "unchanged".

From the last form of (39), we see that if p and ρ are functionally related, i.e. $\rho = \rho(p)$, then $\int_C \dfrac{\nabla p}{\rho} \cdot d\mathbf{r} = \int_{p_0}^{p_0} \dfrac{dp}{\rho(p)} = 0$ so that the circulation is conserved. Another way to see it, is to apply Stokes theorem on the right-hand-side which becomes

$$-\int_C \frac{\nabla p}{\rho} \cdot d\mathbf{r} = \int_S \frac{\nabla \rho \times \nabla p}{\rho^2} \cdot d\mathbf{s}, \tag{40}$$

which is zero if ρ and p are functionally dependent. We now state Kelvin's theorem

Theorem 1 (Kelvin's Theorem) *If the fluid is barotropic, i.e. surfaces of constant p coincide with surfaces of constant ρ then for any closed material contour moving with the fluid*

$$\frac{D\Gamma}{Dt} = 0 \tag{41}$$

The proof of this theorem, like Eq. (37), emphasizes the role of the pressure and density finds in the evolution of vorticity.

One calls the quantity

$$\mathbf{B} = \frac{\nabla \rho \times \nabla p}{\rho^2}, \tag{42}$$

the *baroclinic vector*.

Thus, Kelvin's theorem asserts the conservation of circulation in barotropic, inviscid flows. Flow situations in which $\mathbf{B} \neq 0$ will be considered in the next chapters.

1.4.2   Ertel's Theorem

To probe deeper into the connection between Kelvin's theorem and local variations of vorticity, it is useful to adopt a fully Lagrangian description of the motion, by defining $\mathbf{r}(t; \mathbf{r}^0)$ to be the position (x_1, x_2, x_3) at time t of a fluid particle initially at $\mathbf{r}^0 = (x_1^0, x_2^0, x_3^0)$. Since $\mathbf{r}^0$ labels a particle, it is fixed following the motion and the

Jacobian tensor $J = \{J_{ij}\} = \left\{\dfrac{\partial x_i}{\partial x_j^0}\right\}$ satisfies

$$\frac{D}{Dt} J_{ij} = \frac{\partial}{\partial x_j^0} \frac{Dx_i}{Dt} = \frac{\partial u_i}{\partial x_j^0}$$

$$= \frac{\partial u_i}{\partial x_k} \frac{\partial x_k}{\partial x_j^0} = \frac{\partial u_i}{\partial x_k} J_{kj}. \tag{43a}$$

For any differential $\mathbf{dl}$ separating two nearby points $\mathbf{r}^0$ and $\mathbf{r}^0 + \mathbf{dl}$ on a material curve at time $t = 0$, the quantity $J\mathbf{dl}$ (with components $J_{ij}dl_j$) can be regarded as the vector connecting the two nearby points on the material curve at all subsequent times. By repeating the calculation (43a) for this latter quantity, we obtain

$$\frac{D}{Dt} J_{ij}dl_j = \frac{\partial u_i}{\partial x_k} J_{kj} \; dl_j$$

$$\frac{D}{Dt} (J\mathbf{dl}) = J\mathbf{dl} \cdot \nabla \mathbf{u}. \tag{43b}$$

These calculations remarkably show that ω/ρ in (37) and $\mathbf{dr} \equiv J\mathbf{dl}$ in (43b) satisfy the same equation provided the baroclinic vector $\mathbf{B}$ is zero. Now $\mathbf{dr}$ is a differential along a material curve. The instantaneous "trajectories" of the vector field $\omega/\rho = (\omega/\rho)(\mathbf{r}, t)$ are defined analogously as the integral curves of the system

$$\frac{\mathbf{dr}}{d\sigma} = \frac{\omega}{\rho} (\mathbf{r}(\sigma), t). \tag{44}$$

To understand the meaning of this equation, the vector field ω/ρ is "frozen" (t is fixed), then one starts from an initial position at say $\mathbf{r}(\sigma = \sigma_0)$ and follows the trajectory so that it always stays tangent to the vector field. This procedure defines an instantaneous line or trajectory. We can, at any time, select such an instantaneous trajectory and thereafter follow the evolution of this curve as a material curve. A priori, such a curve as it evolves has no reason to be a solution of (44) anymore. However, since ω/ρ and $\mathbf{dr}$ satisfy the same equation, it follows, assuming uniqueness of the solution, that this material curve remains a trajectory of ω/ρ. Therefore, *the trajectories of ω/ρ are material curves* in a barotropic flow. One usually refers to these trajectories as *vortex lines*, and they can be visualized as being carried about by the fluid although here we consider the vector field ω/ρ instead of the vorticity field alone.

   The first term on the right of (37), which might have the appearance of a "source" of vorticity, is in reality part of the statement of the material nature of vortex *lines*, although ω/ρ itself is not a material invariant. The only true source of vorticity in (37) is the barotropic vector $\mathbf{B}$. The material changes of ω/ρ which occur when $\mathbf{B} = 0$ are due to changes in the geometry of the vortex lines, i.e., to the twisting and stretching of small elements threaded by material lines. As a small pencil of vortex lines, commonly called a *vortex tube* is stretched out by the flow, the cross section diminishes and the local vorticity grows in order to maintain, in accordance with Kelvin's theorem, a fixed circulation about the tube.

The most concise local statement concerning vorticity is an immediate conse-quence of the material nature of vortex lines. If $\mathbf{B} = 0$, the vector $(\boldsymbol{\omega}/\rho)J^{-1} = (\boldsymbol{\omega}/\rho) \cdot \nabla \mathbf{r}^0$ [2] *is a material invariant.*

To see this note first, by differentiating $JJ^{-1} = I$, that

$$\frac{D}{Dt} J^{-1} = -J^{-1} \frac{DJ}{Dt} J^{-1},$$

Thus, using (37) and (43a),

$$\frac{D}{Dt} \left(\frac{\boldsymbol{\omega}}{\rho} J^{-1} \right) = \left[\frac{D}{Dt} \left(\frac{\boldsymbol{\omega}}{\rho} \right) \right] J^{-1} + \frac{\boldsymbol{\omega}}{\rho} \frac{D}{Dt} \left(J^{-1} \right)$$

$$= \left(\frac{\boldsymbol{\omega}}{\rho} \cdot \nabla \mathbf{u} \right) J^{-1} - \frac{\boldsymbol{\omega}}{\rho} J^{-1} (J \cdot \nabla \mathbf{u}) J^{-1} = 0, \qquad (45)$$

which establishes the invariance. Since J is initially the identity tensor, we see that the invariant in (45) is just the initial value of $\boldsymbol{\omega}/\rho$, or

$$\frac{\boldsymbol{\omega}}{\rho} (\mathbf{r}, t) = \frac{\boldsymbol{\omega}}{\rho} (\mathbf{r}^0, 0) J (\mathbf{r}, t). \qquad (46)$$

Now, suppose that λ is any materially invariant scalar, i.e. $D\lambda/Dt = 0$. Then, we may regard λ as a function of the Lagrangian labels $\mathbf{r}^0$, i.e. $\lambda(\mathbf{r}^0) = \lambda(\mathbf{r}) = \text{cst.}$ Then, by remarking that $\nabla \lambda = \nabla_0 \lambda J^{-1}$ and $D(\nabla_0 \lambda)/Dt = 0$, we have

$$\frac{D}{Dt} \left(\frac{\boldsymbol{\omega}}{\rho} \cdot \nabla \lambda \right) = \nabla_0 \lambda \cdot \frac{D}{Dt} \left(\frac{\boldsymbol{\omega}}{\rho} J^{-1} \right) = 0. \qquad (47)$$

We may now state

Theorem 2 (Ertel's Theorem) *Let λ be a scalar field which is invariant by the flow ($D\lambda/Dt = 0$), then provided the fluid is barotropic or λ is a function of ρ and p only, one has*

$$\frac{D}{Dt} \left(\frac{\boldsymbol{\omega}}{\rho} \cdot \nabla \lambda \right) = 0 \qquad (48)$$

---

[2]Or using indices $(\boldsymbol{\omega}/\rho)J^{-1} = \frac{\omega_j}{\rho} \frac{\partial x_i^0}{\partial x_j}$.

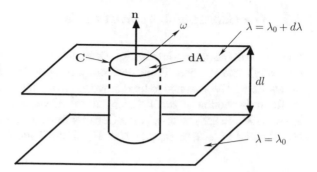

Fig. 3 Kelvin's theorem and Ertel's theorem for a material fluid element of cylindrical shape. Note that the closed contour C always stays on the surface $\lambda = \lambda_0$

The quantity $\frac{\omega}{\rho} \cdot \nabla \lambda$ is called *(Ertel) potential vorticity*. The case where λ is a function of ρ and p only should be clear since $\nabla \lambda = \frac{\lambda}{\rho} \nabla \rho + \frac{\lambda}{p} \nabla p$ is orthogonal to the baroclinic vector **B**. Therefore, in contradistinction to Kelvin's theorem, Ertel's theorem can also be generalized to baroclinic flows. In barotropic flows, $\mathbf{B} = 0$, and λ is unrestricted by being function of ρ and p.

Let us now apply Kelvin's theorem to a small contour C lying on a surface of constant λ, where λ is again a scalar invariant (see Fig. 3).

Kelvin's theorem asserts the constancy of $\omega \cdot \mathbf{dS}$ where $\mathbf{dS} = dS\mathbf{n}$ is the area enclosed by C, while Ertel's theorem asserts the constancy of $(\omega/\rho) \cdot (d\lambda/dl)\mathbf{n}$. Since the mass $\rho dSdl$ is conserved, the two statements become identical in this local setting.

Intuitively, we see that as a tube of vortex lines is laterally compressed, the intensity of the field will increase, by Kelvin's theorem applied to the cross-section of the tube. The presence of ρ in Ertel's local invariant can be traced to the need to simultaneously satisfy Kelvin's theorem and conservation of mass in a small element undergoing deformation. In this sense, Kelvin's theorem is the more fundamental characterization of vorticity. Nevertheless, Ertel's theorem will be useful to us, since in certain cases it is possible to *derive* an invariant λ and to cast the dynamical problem into a form equivalent to Ertel's equation. This is the case, for instance, in the next chapter.

Our vorticity computations so far have been in the inertial frame, but in view of the correspondence

$$\omega_0 = 2\Omega + \omega_r$$

that follows from (22) and (33), one can write Ertel's equation relative to the rotating frame as

$$\frac{d}{dt}(\frac{2\Omega + \omega_r}{\rho} \cdot \nabla \lambda) = 0. \tag{49}$$

1.5   The Geostrophic Approximation

The geostrophic approximation governs flows at small Rossby number, i.e. at relatively rapid rotation. From now on, unless otherwise specified, the equations of motion will be assumed to be relative to a rotating frame, and given by (28).

To study motion at small Rossby number, we shall neglect the inertial terms $\partial_t \mathbf{u} + \mathbf{u} \cdot \nabla \mathbf{u}$ in (28) relative to the Coriolis force. Generally, this requires that both $U/2\Omega L$ and $1/2\Omega T$ be small, where U, L, and T are scales of the velocity field.

1.5.1   The Taylor-Proudman Theorem

The *geostrophic approximation* to (28) is then given by

$$2\Omega \times \mathbf{u} + \frac{1}{\rho}\nabla p = -\nabla \phi_c. \tag{50}$$

The curl of (50) yields

$$-2(\Omega \cdot \nabla)\mathbf{u} + 2\Omega (\nabla \cdot \mathbf{u}) = \mathbf{B}. \tag{51}$$

If the baroclinic vector (42) vanishes, then with $\Omega = (0, 0, \Omega)$ and $\mathbf{u} = (u, v, w)$ the components of (51) reduce to

$$2\Omega \frac{\partial u}{\partial z} = 0, \ 2\Omega \frac{\partial v}{\partial z} = 0, \ 2\Omega \left(\frac{\partial u}{\partial x} + \frac{\partial v}{\partial y} \right) = 0. \tag{52}$$

If, in addition, the fluid can be regarded as incompressible, the constraint $\nabla \cdot \mathbf{u} = 0$ allows us to conclude from (52) that

$$\frac{\partial \mathbf{u}}{\partial z} = 0. \tag{53}$$

This is the *Taylor-Proudman theorem*, stating that the velocity field in the geostropic approximation is invariant along the axis of rotation for a barotropic, inviscid flow. It highlights a striking feature of rotating flows, which was demonstrated experimentally by G. I. Taylor (Fig. 4).

A cylindrical tank of water was rotated with constant speed, and on the bottom of the tank a small cylindrical obstacle was moved horizontally with very small relative speed. In ordinary inviscid fluid dynamics, we would expect the fluid to move over and around the obstacle, the velocity vector being roughly tangent of the boundary. But, on the top of the tank w must vanish everywhere. If fluid particles can only move horizontally, with a velocity independent of z, then the flow field is determined by its structure in a horizontal plane through the cylinder. Indeed, in the experiment, the obstacle was observed to carry an otherwise stagnant column

Fig. 4 The Taylor column,
illustrating the
Taylor-Proudman theorem in
a rapidly rotating fluid

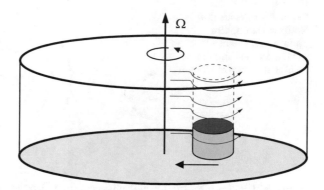

of fluid with it. This "Taylor columns" is an example of the nonlocal effect which
boundary conditions can produce in a rapidly rotating fluid.

1.5.2 Geostrophic Motion

Suppose now that we deal with a rotating layer of fluid of constant density with
$\phi = gz$, and with $w = 0$ rather than $\mathbf{B} = 0$. Then the components of (50) give

$$-2\Omega v = -\frac{1}{\rho}\frac{\partial p}{\partial x}, \quad 2\Omega u = -\frac{1}{\rho}\frac{\partial p}{\partial y}, \quad \frac{\partial p}{\partial z} = -\rho g. \tag{54}$$

This is an example of a *hydrostatic* as well as *geostrophic flow*, wherein the Coriolis
force is in equilibrium with the horizontal pressure gradient.

Looking down on the layer, the flow in the vicinity of a pressure minimum rotates
in the same direction as the large scale rotation. In meteorology, this would roughly
approximate the wind field associated with a low in the pressure field, and is called
a *cyclonic* circulation. Anticyclonic circulations correspond to local highs.

Figure 5 is correct as drawn in the Northern Hemisphere, with the cyclonic
wind blowing counterclockwise around the low. In the Southern Hemisphere, the
cyclonic wind around a low blows clockwise around it. This simple rule is stated
by meteorologists as the Buys-Ballot law: in the Northern Hemisphere, if a person
stands with the back to the wind, atmospheric pressure is low to the left and high to
the right. A more entertaining way of stating it is that, if lost, stand with the back to
wind—if the high pressure is to your left, you're in the Southern Hemisphere.

The geostrophic approximation highlights the unusual features of the dynamics
of a fluid viewed in a rotating coordinate system. To an observer in the inertial frame,
geostrophic flows are predominantly solid body rotation. In order to understand the
physical meaning of the dynamical balance represented in Fig. 5, this background
solid body rotation must be kept in mind. However, it must also be remembered
that a spatial pattern evolving on a moderate time scale relative to a rapidly rotating

Fig. 5 Geostrophic flow in a
Northern Hemisphere
cyclone. Circles are
streamlines, and arrows
indicate the sense of rotation

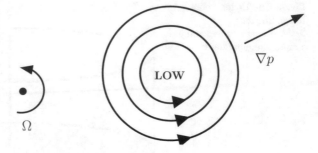

frame will be seen by the inertial observer as a slow modulation of the basic time
dependence due to rotation of the pattern.

It is therefore helpful to deal directly with the relative motion in developing one's
intuition concerning the effects of rotation. For example, in interpreting Fig. 5, the
Coriolis acceleration associated with the eddy is seen to be directed towards its
center, and this acceleration is in equilibrium with minus the pressure gradient.

2 Effects of Shallowness

The most important *geometrical* approximation in geophysical fluid dynamics stems
from the effective shallowness of the fluid layers relative to global horizontal
scales of atmospheric and oceanic flow. In the present chapter, we shall study this
approximation in the simplest setting of an incompressible inviscid fluid of *constant
density*, with uniform body force, namely gravitation, acting vertically downwards.
The resulting approximate description of motion in a frame which rotates about a
vertical axis is usually referred to as rotating shallow-water theory.

The equations of motion for a constant rotation rate Ω are subjected to a
systematic scale analysis, where H is the vertical scale, L is the horizontal scale
and the *aspect ratio* $\delta \equiv H/L$ is small. The shallow-water equations with rotation
are thus derived, and shown to satisfy a particular form of Ertel's theorem. To wit,
potential vorticity $(f_0 + \zeta)/(h - h_0)$ is conserved, where ζ is the fluid's relative
vorticity, $f_0 = 2\Omega$ the background vorticity, and $h - h_0$ the height of a fluid column.

Small-amplitude solutions of these equations are studied in a basin with prescribed depth, using linear theory. Gravity waves are obtained as natural modes
when $f_0 = 0$, and their modification when $f_0 \neq 0$ is investigated, yielding
Poincaré's inertia-gravity waves. The "edge waves" that are related to the Poincaré
waves, and called Kelvin waves, are also derived.

Finally, we study low-frequency modes that satisfy an approximate geostrophic
balance between Coriolis force and pressure-gradient force. Stationary geostrophic
modes, geostrophic contours of a basin and the infinite multiplicity of solutions they
determine are analyzed. Rossby waves over sloping bottom topography are shown

to remove this geostrophic degeneracy, and we examine their relationship with the high-frequency Poincaré and Kelvin waves, along with their physical structure.

2.1 Derivation of the Equations for Shallow Water

For constant $\Omega = (0, 0, \Omega)$ and ρ, we write the governing equations (28) in the form

$$\frac{\partial \mathbf{u}}{\partial t} + (\mathbf{u} \cdot \nabla)\mathbf{u} + 2\Omega \times \mathbf{u} = -\frac{1}{\rho}\nabla p - g\mathbf{i}_3, \tag{55a}$$

$$\nabla \cdot \mathbf{u} = 0, \tag{55b}$$

where $2\Omega = (0, 0, f_0)$, f_0 is the *Coriolis parameter*, and the centripetal acceleration is neglected. We shall think of the fluid layer of constant density ρ, which might represent an ocean basin, as occupying the region $h_0(x, y) \leq z \leq h(x, y, t)$. The upper surface $z = h(x, y, t)$ will be assumed to be a constant pressure p_0 (see Fig. 6). If H is a typical value of h and h_0, and L is a typical horizontal scale of changes in h in p and in the velocity components (u, v, w), the layer is considered shallow if

$$\delta \equiv H/L \ll 1. \tag{56}$$

Since there is no preferred horizontal direction, we take u and v to be of possibly comparable size and of order U.

It is convenient to introduce the horizontal velocity $\mathbf{u}_h = (u, v, 0)$ as well as the horizontal gradient $\nabla_h = (\partial_x, \partial_y, 0)$ and write (55b) in the form

$$\nabla_h \cdot \mathbf{u}_h + \frac{\partial w}{\partial z} = 0. \tag{57}$$

Fig. 6 Shallow layer of fluid

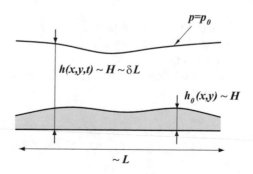

$p = p_0$

$h(x,y,t) \sim H \sim \delta L$

$h_0(x,y) \sim H$

$\sim L$

The first term on the left is then of order U/L. If w varies by an amount of order W over the layer, the second term is of order W/H. These two terms must be comparable in order to account for conservation of matter in the presence of a divergence of horizontal velocity. Thus we take W to be of order δU. This small vertical component of velocity is consistent with small inclination of both surfaces in Fig. 6.

With this scaling, the material derivative is $O(U^2/L)$ and contains no negligible terms, provided that the time scale is taken as L/U. The horizontal pressure forces, which drive the horizontal motions, must be of order comparable to the acceleration. Taking the Coriolis and inertial accelerations to be also comparable to each other, so that the Rossby number $U/2\Omega L$ is $O(1)$, we have $p \sim \rho U^2$. Then the horizontal momentum balance is given by

$$\frac{\partial \mathbf{u}_h}{\partial t} + (\mathbf{u}_h \cdot \nabla_h \mathbf{u}_h + w\frac{\partial \mathbf{u}_h}{\partial z} + 2\Omega \times \mathbf{u}_h = -\frac{1}{\rho}\nabla_h p. \tag{58}$$

On the other hand, the vertical momentum balance is

$$\frac{1}{\rho}\frac{\partial}{\partial z}(p + \rho g z) = O\left(\delta U^2/L\right). \tag{59a}$$

yielding the *hydrostatic approximation*

$$p = -\rho g z + P(x, y, t) + O\left(\delta^2 \rho U^2\right). \tag{59b}$$

For a shallow layer this simplifies therefore to

$$p = -\rho g z + P(x, y, t), \tag{60}$$

where $P(x, y, t)$ is an arbitrary function. The latter is fixed by the dynamic condition $p(x, y, h, t) = p_0 = cst$ on the free surface. Thus

$$p = \rho g(h - z) + p_0. \tag{61}$$

We find that the pressure is also of order $\rho g H \sim \rho U^2$, yielding the characteristic horizontal velocity $U \sim \sqrt{gH}$.

According to (61), $\nabla_h p$ is independent of z. Hence we expect from (58) that $\mathbf{u}_h$ would remain independent of z if initially so. Assuming that this is the case, Eq. (58) simplifies to

$$\frac{D\mathbf{u}_h}{Dt} + 2\Omega \times \mathbf{u}_h + g\nabla_h h = 0, \tag{62a}$$

where, for simplicity, the material derivative in a horizontal plane is denoted by

$$\frac{D}{Dt} \equiv \frac{\partial}{\partial t} + \mathbf{u}_h \cdot \nabla_h \tag{62b}$$

Equation (57) may be integrated to give

$$w \equiv \frac{Dz}{Dt} = -z\nabla_h \cdot \mathbf{u}_h + w_0(x, y, t), \tag{63}$$

involving a second arbitrary function. To determine it, one imposes the kinematic conditions that the upper and lower fluid boundaries be material surfaces. If a boundary is defined implicitly by $S(x, y, z, t) = 0$, the latter condition is that $dS/dt = 0$ on $S = 0$. This gives using (63) applied to $z = h$ and $z = h_0$

$$0 = \frac{D}{Dt}(z - h)_{z=h} = -h\nabla_h \cdot \mathbf{u}_h + w_0 - \frac{Dh}{Dt}, \tag{64a}$$

$$0 = \frac{D}{Dt}(z - h_0)_{z=h_0} = -h_0\nabla_h \cdot \mathbf{u}_h + w_0 - \frac{Dh_0}{Dt} \tag{64b}$$

$$= -h_0\nabla_h \cdot \mathbf{u}_h + w_0 - \mathbf{u}_h \cdot \nabla_h h_0.$$

Thus (64) determines w_0 and provides one equation to supplement (62a). If we subtract one equation from the other, the supplemental equation is obtained in the form

$$\frac{D}{Dt}(h - h_0) = -(h - h_0)\nabla_h \cdot \mathbf{u}_h. \tag{65}$$

Note that (65) has, in horizontal coordinates, the form of a continuity equation for a "density" $h - h_0$ (cf. Eq. (2b)). Physically, such an equation describes how a vertical fluid column changes in height in response to changes in horizontal divergence of the velocity field. Note, in this connection, that the absence of any vertical variation of $\mathbf{u}_h$ implies that vertical columns remain vertical.

The linear variation of w with z in fact implies that the fluid column is stretched uniformly and gives rise to an interesting material invariant of the shallow-water equations. Taking the exact material derivative of $z - h_0$ and using (63) and (64b), we obtain

$$\frac{D}{Dt}(z - h_0) = w - \mathbf{u}_h \cdot \nabla_h h_0 = -z(\nabla_h \cdot \mathbf{u}_h) + w_0 - \mathbf{u}_h \cdot \nabla_h h_0 \tag{66}$$

$$= -(z - h_0)\nabla_h \cdot \mathbf{u}_h.$$

we may thus combine (65) and (66) to obtain

$$\frac{D\lambda}{Dt} = 0, \quad \lambda = \frac{z - h_0}{h - h_0}. \tag{67}$$

Thus, a particle of fluid retains its position in a column as a given fraction of the height of the column.

We are now in a position to apply Ertel's theorem (Sect. 1.4.2) to the material invariant λ. To do this, we need the shallow-water approximation to the inertial or absolute vorticity $\boldsymbol{\omega}_0$. Taking into account that $\mathbf{u}_h$ is independent of z and $(\partial_y w, \partial_x w) = O(\delta U/H)$ we have

$$\boldsymbol{\omega}_0 = \left(\partial_y w - \partial_z v, \partial_z u - \partial_x w, f_0 + \zeta\right) = (0, 0, f_0 + \zeta) + O(\delta\zeta), \tag{68}$$

where $\zeta = \partial_x v - \partial_y u$ is the relative vorticity of the flow. Actually, to justify (68) it must be shown that there is no term of order δ in the equation for $\mathbf{u}_h$, which could contribute $O(1)$ terms through the z-derivatives. But this follows from the estimate (59), which indicates that the shallow-water equations determine the leading terms in an expansion *in* δ^2. Thus, the contributions of $\partial_z u$ and $\partial_z v$ to (68) are in fact of order δ as stated.

Consequently, in the shallow-water approximation, Ertel's theorem, cf. Eq. (49), yields a potential vorticity equation in the form

$$\frac{D}{Dt}\left(\frac{f_0 + \zeta}{h - h_0}\right) = 0, \tag{69a}$$

where $\nabla\lambda = (\nabla_h\lambda, \frac{1}{h - h_0})$ and the term $\nabla_h\lambda$ does not contribute to the scaled gradient $\boldsymbol{\omega}_0\nabla\lambda$ up to order $O(\delta\zeta)$.

We can interpret this equation quite easily in physical terms: since $h - h_0$ is inversely proportional to the horizontal cross-sectional area of a thin column, (69a) is in fact the shallow-water version of a local form of Kelvin's theorem in a rotating frame.

One obtains some simple consequences of (69a) by expanding it as

$$\frac{1}{f_0 + \zeta}\frac{D\zeta}{Dt} = \frac{1}{h - h_0}\frac{D(h - h_0)}{Dt}, \tag{69b}$$

where it is assumed that f_0 does not depend on the horizontal coordinates, i.e. $Df_0/Dt = 0$. Since planetary vorticity f_0 is positive, $f_0 > 0$ in the Northern Hemisphere, it follows from (69b) that relative vorticity ζ and layer depth $h - h_0$ increase and decrease together, at least when ζ is cyclonic, $\zeta > 0$. This monotonic dependence of ζ on $h - h_0$ will also hold for anticyclonic ζ, provided $f_0 \gg |\zeta|$. The latter is typically the case for large-scale geophysical flows. The same conclusions apply in the Southern Hemisphere, where $f_0 < 0$, for cyclonic flows of arbitrary strength and for weak anticyclonic flows; notice that cyclonic flows in the Southern Hemisphere are characterized by $\zeta < 0$, cf. Fig. 3 and accompanying remarks.

Of course, we do not know what the changes in ζ are until we find $h(x, y, t)$. For small Rossby number, relations connecting $\mathbf{u}_h$ (and therefore ζ) to h become quite simple. In this case, (69) becomes a nonlinear partial differential equation in h, which completely determines the shallow-water dynamics.

The shallow-water theory described by Eqs. (62a), (65) has a number of interesting features, both with and without the Coriolis force term. We study here some aspects of the relevant linear theory, and take up in the next chapter some aspects of the nonlinear equations. In other words, the next two sections will address the *small-amplitude* limit of shallow-water theory. Results about the theory's limit for *small Rossby number* can be found in several of the books cited at the beginning of Chap. 1, and they are applied here in Chap. 3.

Remarks It is possible to derive Eqs. (62a), (65) directly from (55), given the hydrostatic approximation (59b) and the fact that the horizontal velocities are independent of z. It suffices to integrate (55) with respect to z, from h_0 to h. The vertically integrated advection term $\int_{h_0}^{h} w \partial_z \mathbf{u}_h \, dz$ is zero. To see this, one uses

$$\int_{h_0}^{h} w \partial_z \mathbf{u}_h dz = [w\mathbf{u}_h]_{h_0}^{h} - \int_{h_0}^{h} \mathbf{u}_h \partial_z w dz = \mathbf{u}_h \frac{D(h - h_0)}{Dt} + \int_{h_0}^{h} \mathbf{u}_h \nabla_h \cdot \mathbf{u}_h dz.$$

Integrating the divergence-free equation (55b), one has $w(z = h) - w(z = h_0) = -(h - h_0)\nabla_h \cdot \mathbf{u}_h$, which yields directly Eq. (65). Plugging this expression into the expression above gives

$$\int_{h_0}^{h} w \partial_z \mathbf{u}_h dz = -\mathbf{u}_h (h - h_0)\nabla_h \cdot \mathbf{u}_h + (h - h_0)\mathbf{u}_h \nabla_h \cdot \mathbf{u}_h = 0.$$

Hence, the horizontal pressure terms, when integrated over the depth of the layer and using (61), yield $\int_{h_0}^{h} (1/\rho)\nabla_h p dz = g(h - h_0)\nabla h$. Therefore, one obtains (62a) multiplied everywhere by the factor $h - h_0$.

Equation (69a) can also be obtained directly from (62a) and (65) by taking the curl to yield

$$\frac{D\zeta}{Dt} + f_0 \nabla_h \cdot \mathbf{u}_h = \frac{D\zeta}{Dt} - f_0 \frac{1}{h - h_0} \frac{D(h - h_0)}{Dt} = (h - h_0)\frac{D}{Dt}\left(\frac{f_0 + \zeta}{h - h_0}\right) = 0.$$

2.2   Reduced-Gravity Shallow-Water Models

In this section, we would like to obtain the so-called $(N + 1/2)$–layer shallow-water model that extends the result above in the direction of a fully 3-D model. The terminology involving a '1/2-layer' is shorthand for the assumption of a deep, motionless layer above which the active flow can be approximated by N layers with nonzero velocities. In this situation, the effective gravity felt in the active layers is smaller than g, and it is referred to as *reduced gravity*. As we shall see at the end of this section, the atmospheric case with $N + 1$ fully active layers is quite similar, except for the need of using a reduced gravity.

More precisely, in the $(N + 1/2)$-layer setting, the oceans are modeled by N active layers of finite depth and constant density ρ_n and another quiescent layer of infinite depth and of larger density ρ_{N+1}. Roughly speaking, the active layers model the ocean above the thermocline and the last layer represents the fluid below it.

The term "reduced gravity" is used because the interfaces below the surface are in slow motion due to the fact that the force acting on them is essentially the gravity field multiplied by a factor that is proportional to density differences between the layers. This factor is in general small, so that the gravity is reduced by an order of magnitude. A simple experiment with a layer of oil above one of water in a large cup can help illustrate this concept: the interface between the two layers, when moving the cup a bit initially, is slowly damped compared to the fast gravity waves at the surface.

In the shallow-water context, the main hypothesis is the validity of the hydrostatic approximation, with constant pressure P_0 at the surface. Since the horizontal gradient of the pressure becomes independent of z, the velocities are also independent of z at all times, if they are initially so; see Sect. 2.1. We assume that the depth associated with each layer is $h_k > 0$, and the vertical coordinate z is taken to be equal to h_0 at the surface, cf. Fig. 7.

Note that the derivatives in the vertical direction must be kept although it is assumed that the velocities $\mathbf{u_k} = (u_k, v_k)$ within each layer k are independent of z. This is due to the fact that the complete velocity field $(\mathbf{u_1}, \cdots, \mathbf{u_N})$ obviously depends on z, although it does so in a discontinuous way, from layer to layer. In particular, when interpreting the derivatives in discretized form, terms like $\partial \mathbf{u_k}/\partial z$ at the interface between layer k and $k + 1$, say, would be proportional to the velocity difference in a first-order linear approximation:

$$A_v \frac{\partial \mathbf{u_k}}{\partial z}\bigg|_{z=h_0-h_1\cdots-h_k} \simeq A_v \frac{\mathbf{u_k} - \mathbf{u_{k+1}}}{h_k + h_{k+1}}. \tag{70}$$

One must be aware that estimating these derivatives is more a matter of parametrization of subgrid-scale processes, and it is difficult to assess what they should be in reality. For oceanic applications, these vertical derivatives are either

Fig. 7 Reduced-gravity shallow-water model with N active layers of density ρ_k and thickness $h_k > 0$, while the bottom layer $N + 1$ is of infinite depth and at rest. The surface deviation is $z = h_0$ and serves as the reference for the vertical coordinate. The pressure P_0 at the surface is constant

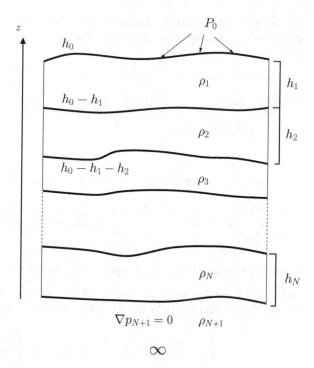

neglected, except at the surface, or take the form of Rayleigh damping between the layers, in view of (70).

We thus start from N 2-D coupled Navier-Stokes equations in dimensional form, that is

$$\frac{\partial \mathbf{u_k}}{\partial t} + (\mathbf{u_k} \cdot \nabla)\mathbf{u_k} + w_k \frac{\partial \mathbf{u_k}}{\partial z} + 2\mathbf{f} \times \mathbf{u_k} = -\frac{1}{\rho_k}\nabla p_k + A_H \Delta \mathbf{u_k} + A_v \frac{\partial^2 \mathbf{u_k}}{\partial z^2},$$
$$\nabla \cdot \mathbf{u_k} + \frac{\partial w_k}{\partial z} = 0.$$

$$(71)$$

Using the hydrostatic hypothesis, and the continuity of the pressure at each interface, one obtains the pressure within each layer:

$$p_k(z) = P_0 + \sum_{j=1}^{k-1} g\rho_j h_j + g\rho_k \left(h_0 - z - \sum_{j=1}^{k-1} h_j \right), \quad 1 \le k \le N + 1. \qquad (72)$$

In order to proceed, we reformulate the nonlinear terms in (73) in a more convenient conservative form using the divergence-free equation, to obtain

$$
\frac{\partial u_k}{\partial t} + \nabla \cdot (u_k \mathbf{u_k}) + \frac{\partial (u_k w_k)}{\partial z} - 2 f v_k = -\frac{1}{\rho_k} \frac{\partial p_k}{\partial x} + A_H \Delta u_k + A_v \frac{\partial^2 u_k}{\partial z^2},
$$
$$
\frac{\partial v_k}{\partial t} + \nabla \cdot (v_k \mathbf{u_k}) + \frac{\partial (v_k w_k)}{\partial z} + 2 f u_k = -\frac{1}{\rho_k} \frac{\partial p_k}{\partial y} + A_H \Delta v_k + A_v \frac{\partial^2 v_k}{\partial z^2},
$$
$$
\nabla \cdot \mathbf{u_k} + \frac{\partial w_k}{\partial z} = 0.
$$

$$(73)$$

In the following, we will extensively use the formula

$$
\frac{d}{dx} \int_{a(x)}^{b(x)} f(u, x)\, du = \int_{a(x)}^{b(x)} \frac{\partial f}{\partial x}(u, x)\, du + \frac{db}{dx} f(b(x), x) - \frac{da}{dx} f(a(x), x).
$$

We now integrate (73) over the layer depth, that is between $z^- = h_0 - \sum_{j=1}^{k} h_j$ and $z^+ = z^- + h_k$, where z^- and z^+ of course depends on (x, y, t). We proceed with each term and, again assuming that $\mathbf{u_k}$ does not depend on z, obtain:

$$
\int_{z^-}^{z^+} \frac{\partial \mathbf{u_k}}{\partial t}\, dz = \frac{\partial}{\partial t} \int_{z^-}^{z^+} \mathbf{u_k}\, dz + \mathbf{u_k} \left(\frac{\partial z^-}{\partial t} - \frac{\partial z^+}{\partial t} \right)
$$
$$
= \frac{\partial (\mathbf{u_k} h_k)}{\partial t} - \mathbf{u_k} \frac{\partial h_k}{\partial t}.
$$

$$
\int_{z^-}^{z^+} \frac{\partial (u_k w_k)}{\partial z}\, dz = u_k \left(w_k(z^+) - w_k(z^-) \right) = u_k \left(\frac{Dz^+}{Dt} - \frac{Dz^-}{Dt} \right) = u_k \frac{Dh_k}{Dt}.
$$

$$
\int_{z^-}^{z^+} -\frac{1}{\rho_k} \nabla p_k\, dz = -g \sum_{j=1}^{k-1} \left(\frac{\rho_j - \rho_k}{\rho_k} \right) h_k \nabla h_j - g h_k \nabla h_0,
$$

$$
\int_{z^-}^{z^+} \nabla \cdot (u_k \mathbf{u_k})\, dz = \nabla \cdot \int_{z^-}^{z^+} u_k \mathbf{u_k}\, dz - \left(\frac{\partial z^+}{\partial x} - \frac{\partial z^-}{\partial x} \right) u_k^2 - \left(\frac{\partial z^+}{\partial y} - \frac{\partial z^-}{\partial y} \right) u_k v_k
$$
$$
= \nabla \cdot (u_k h_k \mathbf{u_k}) - (u_k \mathbf{u_k} \cdot \nabla) h_k.
$$

using the relation (72).

$$
\int_{z^-}^{z^+} \left(\nabla \cdot \mathbf{u_k} + \frac{\partial w_k}{\partial z} \right) dz = w_k(z^+) - w_k(z^-) + \nabla \cdot \int_{z^-}^{z^+} \mathbf{u_k}\, dz - (\mathbf{u_k} \cdot \nabla)(z^+ - z^-)
$$
$$
= \frac{Dh_k}{Dt} + \nabla \cdot (\mathbf{u_k} h_k) - (\mathbf{u_k} \cdot \nabla) h_k = \frac{\partial h_k}{\partial t} + \nabla \cdot (\mathbf{u_k} h_k).
$$

$$
\int_{z^-}^{z^+} \Delta u_k\, dz = h_k \Delta u_k = \Delta(h_k u_k) - [2\nabla h_k \cdot \nabla u_k + u_k \Delta h_k].
$$

To complete the derivation, one integrates the vertical dissipative terms:

$$\int_{z^-}^{z^+} A_v \frac{\partial^2 \mathbf{u_k}}{\partial z^2}\, dz = A_v \frac{\partial \mathbf{u_k}}{\partial z}\bigg|_{z^+} - A_v \frac{\partial \mathbf{u_k}}{\partial z}\bigg|_{z^-} = \tau^{z,k}(z^+) - \tau^{z,k}(z^-).$$

The terms on the right-hand side of the last equation above correspond to the horizontal stresses acting on the top and bottom of the layer k. In particular, the stress at the surface is the wind stress, see Fig. 15 and (101) in Chap. 3.

In order to conclude, one must relate h_0 to the layer thicknesses. The requisite relation is readily provided by the motionless $N + 1$-st layer. Absence of motion implies that there are no pressure variations, that is

$$\nabla p_{N+1} = 0. \tag{74}$$

In the following sections, we will use the notation

$$g_j^i = g \frac{\rho_j - \rho_i}{\rho_j}. \tag{75}$$

We thus find, using (72) and (74), that

$$g \nabla h_0 = \sum_{j=1}^{N} g_{N+1}^{j} \nabla h_j, \tag{76}$$

We are now ready to write the reduced-gravity shallow-water model with N active and one passive layer, using the mass flux vector notation $\mathbf{U_k} := \mathbf{u_k} h_k$ and the fact that $\frac{D h_k}{Dt} = \frac{\partial h_k}{\partial t} + (\mathbf{u_k} \cdot \nabla) h_k$.

$$\frac{\partial \mathbf{U_k}}{\partial t} + \nabla \cdot (\mathbf{U_k} \otimes \mathbf{u_k}) + \mathbf{f} \times \mathbf{U_k} = -h_k \mathcal{G} \cdot \nabla \mathcal{H} + h_k (\mathcal{R} \nabla \mathcal{H})_k + \mathcal{D}_H + \mathcal{T}_k$$

$$\frac{\partial h_k}{\partial t} + \nabla \cdot \mathbf{U_k} = 0, \tag{77}$$

Here $\nabla \cdot (\mathbf{U_k} \otimes \mathbf{u_k})$ is a convenient notation for $(\nabla \cdot (U_k \mathbf{u_k}), \nabla \cdot (V_k \mathbf{u_k}))$,

$$\mathcal{T}_k = \tau^{z,k}(z^+) - \tau^{z,k}(z^-),$$

$\mathcal{H} = (h_1, \cdots h_N)$, $\nabla\mathcal{H} = (\nabla h_1, \cdots \nabla h_N)$, $\mathcal{R}$ is a $N \times N$ matrix given by

$$
\mathcal{R} = \begin{pmatrix}
0 & 0 & \cdots & \cdots & \cdots & \cdots & 0 \\
g_2^1 & 0 & \cdots & \cdots & \cdots & \cdots & 0 \\
g_3^1 & g_3^2 & 0 & \cdots & \cdots & \cdots & 0 \\
\vdots & & & & & & \\
g_k^1 & \cdots & g_k^{k-2} & g_k^{k-1} & 0 & \cdots & 0 \\
\vdots & & & & & & \\
g_N^1 & \cdots & \cdots & \cdots & \cdots & g_N^{N-1} & 0
\end{pmatrix}
\tag{78}
$$

and the surface deviation in the densities that corresponds to h_0 in vector form is $\mathcal{G} = (g_{N+1}^1, \cdots, g_{N+1}^N)$. The interfacial stress $\mathcal{T}_k$ for $2 \leq k \leq N - 1$ is often assumed to be zero to retain only the surface and bottom stress. Since we assume the velocity vanishes in the $N + 1$-st layer, one expects to get a boundary layer of Ekman type.

Herewith we just obtained the shallow-water equations in a conservative mass-flux form that is quite convenient for numerical discretization. There are two other ways to express (77), one is the momentum form and the other is the vorticity form. In momentum form, the equations are indeed straightforward to obtain since it suffices to divide by h_k and use the mass-conservation equation:

$$
\frac{\partial \mathbf{u_k}}{\partial t} + (\mathbf{u_k} \cdot \nabla)\mathbf{u_k} + \mathbf{f} \times \mathbf{u_k} = -\mathcal{G} \cdot \nabla\mathcal{H} + (\mathcal{R}\nabla\mathcal{H})_k + \frac{\mathcal{D}_H}{h_k} + \frac{\mathcal{T}_k}{h_k}
$$
$$
\frac{\partial h_k}{\partial t} + \nabla \cdot \mathbf{U_k} = 0.
\tag{79}
$$

The vorticity formulation is found by cross-differentiating (79) and defining the vorticities, layer-by-layer, as $\zeta_k = \frac{\partial v_k}{\partial x} - \frac{\partial u_k}{\partial y}$. Mass conservation implies that $\nabla \cdot \mathbf{u_k} = -\frac{1}{h_k}\frac{D_k}{Dt}h_k$ and the cross-differentiation of the left-hand side of (79) yields the expression $\frac{D_k}{Dt}(\zeta_k + f) + (f + \zeta_k)\nabla \cdot \mathbf{u_k}$. Thus, the final result is

$$
h_k \frac{D_k}{Dt}\left(\frac{\zeta_k + f}{h_k}\right) = \nabla \times \frac{\mathcal{D}_H}{h_k} + \nabla \times \frac{\mathcal{T}_k}{h_k}.
\tag{80}
$$

Equation (80) neatly generalizes the expression (69).

Note that the potential vorticity $(\zeta_k + f)/h_k$ in each layer is a conservative quantity in the absence of lateral friction and vertical stresses. Of course, one needs to relate h_k with ζ_k to obtain a complete description of the flow. It therefore amounts to solve the shallow-water equations in their primitive form, that is (77) or (79). In the quasi-geostrophic approximation, one can nevertheless find a closed equation for the vorticity–streamfunction variables. Again, we refer, for instance, to Chapter 3 in [56] for the derivation and to Chap. 3 for an application to the wind-driven ocean circulation.

Remark The case of a genuine $N + 1$-layer model is slightly different. In such a model, the relation (72) for the pressure in each layer is the same but instead of having (76) due to $\nabla p_{N+1} = 0$, one has $g\nabla h_0 = g\nabla h_1 + \cdots + g\nabla h_{N+1}$, which involves an additional layer $N + 1$.

2.3 Small-Amplitude Motions in a Basin

From now on, we will omit the subscript 'h', namely ∇ and $\mathbf{u}$ are understood as ∇_h and $\mathbf{u}_h$. Suppose that the fluid lies in a basin occupying a closed set $\mathcal{D}$ in the (x, y)-plane, cf. Fig. 8, and that when at rest, i.e. with $h = \text{const.}$, the depth of the fluid is positive, $H(x, y) > 0$, over $\mathcal{D}$. We then set

$$h - h_0 = H(x, y) + \eta(x, y, t), \tag{81}$$

where η and hence $\mathbf{u} = (u, v)$ are small perturbations. Since nonlinear terms in the perturbations are neglected, (62a) and (65) reduce to

$$\frac{\partial u}{\partial t} - f_0 v + g\frac{\partial \eta}{\partial x} = 0, \tag{82a}$$

$$\frac{\partial v}{\partial t} + f_0 u + g\frac{\partial \eta}{\partial y} = 0, \tag{82b}$$

$$\frac{\partial \eta}{\partial t} + \nabla \cdot (H\mathbf{u}) = 0. \tag{82c}$$

When $f_0 = 0$, we obtain the equations for small amplitude oscillations in a nonrotating basin. In this case (82) can be combined to give

$$\frac{\partial^2 \eta}{\partial t^2} - g\nabla \cdot (H\nabla \eta) = 0 \tag{83a}$$

If the contour $\partial \mathcal{D}$ is an impenetrable barrier, the appropriate boundary condition for system (82) is $\mathbf{u} \cdot \mathbf{n} = 0$, where $\mathbf{n}$ is the normal to $\partial \mathcal{D}$. In this case, (82a), (82b) with

Fig. 8 A shallow-water basin $\mathcal{D}$, with layer thickness $h - h_0 = H(x, y) + \eta(x, y, t)$. The deviation η is assumed to be small compared with H

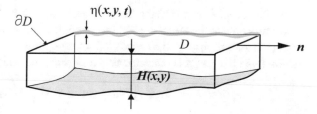

$f_0 = 0$ yields

$$\nabla\eta \cdot \mathbf{n} = 0 \tag{83b}$$

as the corresponding boundary condition for (83a).

If we look for eigenfunctions $\eta(x, y, t) = e^{i\omega t} N(x, y)$ of (83), where N has zero integral over $\mathcal{D}$, then the resulting equation for N is equivalent to the Euler-Lagrange equation for the following variational problem:

$$\min_N \int_{\mathcal{D}} H|\nabla N|^2 dxdy, \quad \int_{\mathcal{D}} N^2 dxdy = 1.$$

Here the minimum is to be taken over all N which satisfy $\partial N/\partial\mathbf{n} = 0$ on the boundary and have zero integral over $\mathcal{D}$. The Lagrange multiplier in the problem may then be identified with ω^2/g. In this way, the minimum frequency ω_0 is determined variationally by

$$\omega_0^2 = g \min_N \frac{\displaystyle\int_{\mathcal{D}} H|\nabla N|^2 dxdy}{\displaystyle\int_{\mathcal{D}} N^2 dxdy}.$$

We note in passing that, for given $H(x, y)$, ω_0 is strictly positive and the Rayleigh-Ritz procedure provides an effective means for constructing approximate eigenfunctions. These eigenfunctions represent standing gravity waves in a shallow basin.

The relative simplicity of this theory is a consequence of the self-adjoint property of the boundary-value problem (83). Writing (83a) as

$$\frac{\partial^2\eta}{\partial t^2} = L\eta,$$

it is easily seen that L is formally self-adjoint,

$$\int_{\mathcal{D}} \phi L\psi \, dxdy = \int_{\mathcal{D}} \psi L\phi \, dxdy,$$

for all functions ϕ, ψ satisfying the boundary condition (83b).

We turn now to the effect of rotation on small-amplitude motions. It can be shown from (82) that, if $f_0 \neq 0$, the associated linear operator is no longer self-adjoint. To exhibit this operator, it is convenient to introduce the quantities

$$\sigma = \nabla \cdot (H\mathbf{u}), \text{ and } s = \nabla \times (H\mathbf{u}) \tag{84}$$

which may be thought of as representing the divergence and vorticity of the integrated mass flux across the layer. From (82a), (82b), we obtain by differentiation

$$\frac{\partial \sigma}{\partial t} - f_0 s + g\nabla \cdot (H\nabla\eta) = 0, \tag{85a}$$

$$\frac{\partial s}{\partial t} + f_0\sigma + g\nabla H \times \nabla\eta = 0 \tag{85b}$$

Eliminating s and using $\sigma = -\frac{\partial \eta}{\partial t}$ from (84) and (82c), the equation for η becomes

$$\frac{\partial}{\partial t}\left[\frac{\partial^2 \eta}{\partial t^2} + f_0^2\eta - g\nabla \cdot (H\nabla\eta)\right] = f_0 g\nabla H \times \nabla\eta \tag{86}$$

We consider the simplest case, namely H = constant, so that the last two terms of the above equation vanish. One substitutes $\eta = e^{i(\omega t + \mathbf{k}\cdot\mathbf{x})}$ into (86) to obtain a dispersion equation for traveling plane waves in an unbounded domain:

$$i\omega[f_0^2 - \omega^2] = -i\omega g H|\mathbf{k}|^2, \tag{87}$$

where the wavenumber $\mathbf{k}$ has two components (k_x, k_y) and $|\mathbf{k}|^2 = k_x^2 + k_y^2$. Disregarding for the moment the wave corresponding to $\omega = 0$, these waves are known as Poincaré or *inertia-gravity waves*, and represent the generalization to the rotating case of familiar gravity waves. For every angular frequency $\omega \neq 0$, there are two waves with phase speed along the direction $\mathbf{k}$

$$|c_\mathbf{k}| = -\omega/|\mathbf{k}| = \pm\left(\frac{f_0^2}{|\mathbf{k}|^2} + gH\right)^{1/2}. \tag{88}$$

One propagates in the positive, the other in the negative direction relative to the wave vector $\mathbf{k}$. They are observed in the atmosphere as well as in large bodies of water.

Seen from above, particle paths in the nonrotating case appear as line segments parallel to the direction of propagation of the wave; see panel (a) in Fig. 9. With rotation, cf. panel (b), the segments become ellipses whose eccentricity decreases with rotation. These facts can be most easily derived from a Lagrangian description.

For Poincaré waves in a finite basin, one finds that, in certain simple basin shapes, the boundary condition selects a discrete set of eigenfunctions. Each of the latter exhibits a dependence of frequency upon a discrete ordering parameter that is similar to (87).

Using (87), we write the dispersion relation for inertia-gravity waves in the form

$$\left(\frac{\omega}{f_0}\right)^2 = 1 + L_R^2|\mathbf{k}|^2, \tag{89}$$

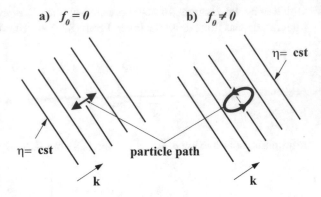

Fig. 9 Wave fronts, which correspond to $\eta = $ const., and particle paths for gravity waves: (**a**) without rotation; and (**b**) with rotation

where

$$L_R = \frac{\sqrt{gH}}{f_0} \tag{90}$$

has the dimensions of a length, and is called the *external Rossby radius of deformation*. The physical importance of this length will become clearer in studying the near-surface ocean circulation in Sect. 3.

In the present context, $2\pi L_R$ is a horizontal scale below which the Poincaré waves become essentially gravity waves with small Coriolis effect. To take a typical case for the Earth's atmosphere, let $H = 10$ km, giving $L_R \cong 3 \times 10^3$ km. Thus, for horizontal scales of $2\pi/k < 3 \times 10^3$ km, we note that the gravity contribution to (89) dominates.

Notice that the frequency of these waves increases with wavenumber **k**. It is always larger than f_0, i.e., their period is smaller than the period of planetary rotation. A typical period of the inertia-gravity waves at length scales $L_R = O\left(10^3 \text{ km}\right)$ is about 2 h. For most geophysical problems on such horizontal scales, this determines a short time scale. An important feature, therefore, of models for large-scale atmospheric and oceanic dynamics is an appropriate mechanism for filtering out the possible Poincaré-wave components of small period. Inertia-gravity waves are observed to have, in fact, small amplitudes over these length scales.

2.3.1 Kelvin Waves

Another possible component of oscillations of a basin is the family of Kelvin waves. These waves can be generated near a bounding contour and oscillate at a frequency which can be close to that associated with gravity waves. To take the simplest case, suppose that the basin is in the upper half-plane $y > 0$.

Eliminating u from (82a), (82b) we obtain

$$\frac{\partial^2 v}{\partial t^2} + f_0^2 v + g\frac{\partial^2 \eta}{\partial y \partial t} - f_0 g\frac{\partial \eta}{\partial x} = 0. \tag{91}$$

if v vanishes on the boundary we must also have

$$\frac{\partial^2 \eta}{\partial y \partial t} - f_0\frac{\partial \eta}{\partial x} = 0, \quad y = 0. \tag{92}$$

With $\eta = N(y)e^{i(\omega t + k_x x)}$, $k_x > 0$, Eq. (86) with $H = \text{cst}$ yields

$$gH\frac{d^2 N}{dy^2} - \left(f_0^2 - \omega^2 + gHk_x^2\right)N = 0, \, y > 0 \tag{93a}$$

$$\omega\frac{dN}{dy} - k_x f_0 N = 0, \quad y = 0. \tag{93b}$$

A new result can be obtained if we assume $f_0^2 - \omega^2 + gHk_x^2 > 0$, since then the y-dependence involves real exponentials rather than real trigonometric functions. To obtain a solution bounded in the upper half plane we must select the decaying exponential. Then (93b) requires $N = N_0 e^{(f_0 k_x/\omega)y}$, $\omega < 0$, in which case (93a) gives

$$\left(\omega^2 - gHk_x^2\right)\left(f_0^2 - 1\right) = 0. \tag{94}$$

Taking $\omega = -k_x\sqrt{gH}$, one obtains a wave with positive phase velocity identical to that of a pure gravity wave.

It can be checked from (82) that v vanishes identically for this wave, and that wave crests are parallel to the y-axis. On each line $y = \text{cst}$ the wave propagates as an ordinary gravity wave, but the x-motion is in geostrophic balance with the wave height η; the latter decreases in concert with u as y increases, at an exponential rate determined by the Rossby radius L_R.

Because of their exponential decrease away from the boundary, Kelvin waves can be regarded as "edge waves". They have vanishing frequency as $k_x \to 0$, a property which separates them from inertial-gravity waves; see Fig. 12 in section below. But for moderate length scales in the x-direction they fall into the category of "fast" waves. Kelvin waves are particularly important in tropical meteorology and oceanography, where the equator plays the role of the "wall".

2.4 Geostrophic Degeneracy and Rossby Waves

The time scales associated with large-scale atmospheric phenomena are of the order
of days, while those for the ocean are of the order of months. If these time scales are
to be reflected in the small-amplitude solutions of the shallow-water equations, the
relevant physics must be associated with the modes corresponding to the root $\omega = 0$
of the dispersion relation (87) in the constant-depth case.

2.4.1 Stationary Geostrophic Modes

Let us consider then the stationary modes in the small-amplitude theory. Equations (82), reduce for steady solutions to

$$2\Omega \times \mathbf{u}_h + g\nabla_h \eta = 0 \tag{95a}$$

$$\nabla_h \cdot (H\mathbf{u}_h) = 0 \tag{95b}$$

   This is a special kind of geostrophic flow (Sect. 1), and we refer to all solutions
of (95) compatible with the condition on the boundary of the basin as stationary
geostrophic modes. Equation (95a) implies that $(g/f_0)\,\eta$, is a stream function for the
flow's velocity field and that lines $\eta = $ cst are streamlines of the flow. In particular,
the boundary has to be an isoline of η.
   For the constant-depth problem we see that arbitrary divergence-free horizontal
motions compatible with the condition that the boundary be a streamline are
allowed. Equations (95) simply tells us what the fluid level must be to bring
the system into geostrophic balance. From (86) and (95) we conclude that in
the constant-depth case, with η proportional to $e^{i\omega t}$ in separated variables, the
eigenvalue $\omega = 0$ is infinitely degenerate. Given any boundary contour, there are
an infinite number of functions η which satisfy the boundary conditions.
   If H depends upon x and y, the geostrophic modes may still be infinitely
degenerate, since (95b) then implies that the streamfunction η may be an arbitrary
function of H alone. Thus lines of constant depth and streamlines coincide.
Although contours of constant H that intersect the boundary ∂D must carry zero
velocity, closed contours that do not touch the boundary—or nested islands of such
contours—do allow the construction of geostrophic modes, as shown in Fig. 10).
The latter type of contours, along which stationary geostrophic flow may occur, are
often called geostrophic contours.

2.4.2 Sloping Bottom Topography

The question now arises as to what happens to the "disappearing" geostrophic
modes when the depth function is perturbed, in such a way that the geostrophic

Fig. 10 Geostrophic contours for small-amplitude motion in a basin $\mathcal{D}$

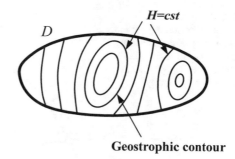

Geostrophic contour

degeneracy is partially or perhaps completely removed. To take a particularly simple case, consider a large basin whose depth function is locally linear

$$H = H_0 \left(1 - \beta y\right), \quad \beta > 0. \tag{96}$$

It is helpful to consider the parameter β as small, $\beta = O(\varepsilon)$, to retain (96) over a sizeable region, and to disregard boundary effects.

It is customary in dealing with large-scale flow of the atmosphere and oceans to choose the x-coordinate pointing eastward and the y-coordinate pointing northward. Such a geometry is able to simulate the important effect of latitudinal variation in the effective local rate of rotation, as shown, for instance by Weeks and colleagues [169, and references therein].

With (96), the geostrophic contours are the lines $y = $ const., so the degeneracy of the constant-depth case has been significantly altered. To force recovery of the missing modes in (82) as a perturbation, we suppress completely the remaining geostrophic modes by requiring that η be of the form

$$\eta = \tilde{\eta} e^{i(\omega t + k_x x)}, \tag{97a}$$

with similar expressions for u and v, and with small ω/f_0.

The smallness of ω guarantees slow changes in all the variables; in particular, we shall have small $\partial u/\partial t$ and $\partial \eta/\partial t$. Equations (82a), (82c) then imply an approximate *geostrophic balance*. Since $\partial \eta/\partial y = 0$, (82b) also states that u itself is small. The solutions we consider have therefore predominantly y-directed fluid velocity. That is, particle motion of the slowly-varying flow is essentially perpendicular to the geostrophic contours of the stationary flow discussed before, while the slow wave propagation is parallel to these contours. Moreover, the free surface η and velocity (u, v) of the flow (97a) are still in a nearly geostrophic, slowly shifting balance.

Substituting η from (97a) and H from (96) into (86) yields

$$i\omega \left(f_0^2 - \omega^2\right) = -gH_0 i\omega \left(1 - \beta y\right) k_x^2 + igH_0 \beta f_0 g k_x. \tag{97b}$$

Fig. 11 Linear wave
solutions for sloping bottom
topography. The quantities
LHS and RHS refer to the
left- and right-hand side of
Eq. (98), respectively

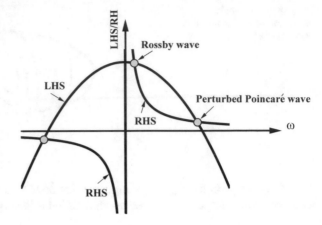

which involves y explicitly in the first term on the right. This appears to be inconsistent with the assumed form of η. However, we are dealing only with small ω, and one can thus neglect the term βy in order to obtain a dispersion relation that is still correct to $O(\varepsilon)$. Rearranging terms, the latter takes the form

$$f_0^2 + g H_0 k_x^2 - \omega^2 = \frac{g H_0 \beta f_0 k_x}{\omega}. \tag{98}$$

The solutions of (98) are plotted in Fig. 11, which shows that, for small β, there are two "perturbed" inertia-gravity modes, together with a third root of order β. As a matter of fact, letting the left-hand side (LHS) vanish gives directly (89). Since this third root is small, an approximate dispersion relation for it is

$$\omega = \frac{g H_0 f_0 k_x}{f_0^2 + g H_0 k_x^2} \beta. \tag{99}$$

The waves associated with (99) are called *Rossby waves* and, in the present case of positive β, their phase velocity $c = -\omega/k_x$ is westward. The frequency of these waves is always smaller than f_0, provided β is not too large. Rossby waves are further distinguished from both Poincaré and Kelvin waves by the fact that, for short waves, ω decreases with k_x, as seen in Fig. 12.

We might think of these waves as replacing, among the infinity of geostrophic modes in a basin of constant depth, the family of modes with $y = $ const. as streamlines. In fact, to $O(\varepsilon)$, the flow is still nondivergent and in geostrophic balance,

$$u = -\frac{\partial \psi}{\partial y}, \quad v = \frac{\partial \psi}{\partial x}, \tag{100a}$$

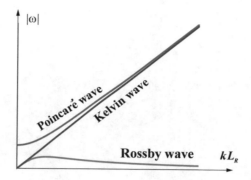

Fig. 12 Comparison of dispersion relations for small-amplitude shallow-water waves with rotation, namely $\omega_{\text{Poincaré}} = \pm f_0(1 + x^2)^{1/2}$ and $\omega_{\text{Kelvin}} = \pm f_0 x$, and for Rossby waves, $\omega_{\text{Rossby}} = \beta(gH)^{1/2}x/(1 + x^2)$, where $x = kL_R$. The ordinate corresponds to the absolute value $|\omega|$ of the frequency and the abscissa is in units of the Rossby radius L_R, cf. (90)

where $\psi = (g/f_0)\eta$. This balance also implies that, to $O(\varepsilon)$, the relative vorticity in the wave, cf. (68), is given by

$$\zeta = \nabla^2 \psi. \tag{100b}$$

But it is precisely the small deviation from exact geostrophy—apparent to $O(1)$ only in the vorticity form of the equations, cf. (86), (97)—that eliminates the geostrophic degeneracy and gives rise to the waves.

To understand the physical structure of Rossby waves, it is useful to consider their propagation in terms of vorticity balance. According to (69), vorticity and depth along a material trajectory increase and decrease together. Examining the structure of the Rossby wave (97a) at any given time, we observe that ζ and v have the appropriate variations shown in Fig. 13, namely if $\zeta(x) = \sin k_x x$ then $v(x) = k_x \cos k_x x$. Since v advects ζ, wherever v is positive the relative vorticity at that point must be instantaneously decreasing with time, since advection would carry the fluid element from a deeper region into a shallower region ($\beta > 0$). In order for this to happen, the wave pattern must drift westward.

3 The Wind-Driven Ocean Circulation

3.1 Introduction and Motivation

Having introduced the reader to some of the fundamentals of planetary-scale flows in Chaps. 1 and 2, we are ready to turn to current topics of the climate sciences and of the applications of dynamical systems theory to them.

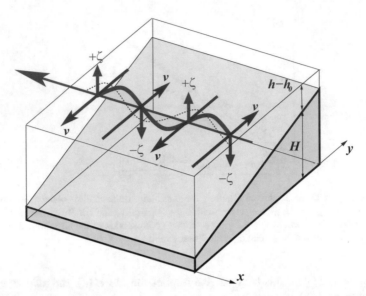

Fig. 13 Linear Rossby waves must propagate westward when $H = H_0(1 - \beta y)$, with $\beta > 0$, in order to conserve potential vorticity according to (69a). Thus, an initial vorticity wave given by (97a), plotted in the figure as the heavy blue undulating contour, has its vorticity decreasing—as given by the vertical red arrow that indicates the vorticity contribution $-\zeta$ in this case—if the particle moves northward. To the contrary, southward particle motion makes a positive contribution $+\zeta$ to the wave's vorticity

Since the early 1970s, the importance of the climate sciences has been rapidly growing, as a field of scholarship as well as a source of information for stewardship of planet Earth [71–74, 114]. Likewise, the usefulness of dynamical systems and of their successive bifurcations in navigating across the hierarchy of climate models [37, 53, 54, 56, 136]—while far from universally accepted—has found a wide community of users and produced a substantial literature.

An important recent development in this literature was caused by the realization that the theory of differentiable dynamical systems (DDS)—in its classical form dealing with *autonomous* DDSs—was fairly well adapted to applications in which neither the forcing nor the coefficients were time dependent. To a good approximation, this could be said to be the case in numerical weather prediction [79, 159], but much less so in climate problems in which the seasonal forcing [18, 59, 65, 77, 78, 135, 162, 163] or anthropogenic effects [20, 71–73, 114] play a major role.

To cope with the latter type of problems, applications of the theory of *nonautonomous and random dynamical systems* (NDS and RDS) have started to appear in the theoretical climate dynamics literature [10, 23, 43, 64]. The purpose of the present paper is to give some flavor of the systematic application of DDS, NDS and RDS theory across the hierarchy of climate models, from the simplest, conceptual ones all the way to the most elaborate ones [39, 53, 54, 59].

In mathematical terms, this means considering, at the lower end of the hierarchy, small systems of ordinary differential equations (ODEs) and at the higher end full-blown systems of nonlinear partial differential equations (PDEs). In the climate literature, the former are sometimes called "toy models," while the latter are referred to as general circulation models (GCMs) or, more recently, as global climate models, which preserves the GCM acronym.

An interesting problem to which these ideas and results will be applied herein is that of the wind-driven circulation in mid-latitude ocean basins [37, 39], subject at first to time-constant and then to purely periodic and more general forms of time-dependent wind stress. The wind-driven circulation is obviously strongest near the surface of the oceans, and its variability lies mostly in the subannual and interannual range. In many simplified, theoretical treatments of the oceans, this circulation is treated separately from the so-called buoyancy-driven, thermohaline or overturning circulation. The latter is driven by mass fluxes at the ocean–atmosphere interface; it involves the density distribution of water masses—which depends on both temperature and salinity, hence its characterization as thermohaline—and it reaches all the way to the bottom of the oceans. The variability of this overturning circulation lies largely in the interdecadal and centennial range [37, 39, 52, 54, 115, 121, 153, 157]. It will only be mentioned here in passing in Sect. 3.4.

Key features of the wind-driven circulation are described in Sect. 3.2 below. The influence of strong thermal fronts—like the Gulf Stream in the North Atlantic or the Kuroshio in the North Pacific—on the mid-latitude atmosphere above is severely underestimated, thus contributing to the current range of uncertainties in climate simulations over the twenty-first century. Typical spatial resolutions in recent century-scale GCM simulations [1, 71–73, 128, 151] are still of the order of 100 km at best, whereas resolutions of 50 km and less would be needed to really capture the strong mid-latitude ocean-atmosphere coupling just above the oceanic fronts [47, 48, 108, 148].

An important additional source of uncertainty comes from the difficulty to correctly parametrize global and regional effects of clouds and their highly complex small-scale physics. This difficulty is particularly critical in the tropics, where large-scale features such as the El-Niño/Southern Oscillation (ENSO) and the Madden-Julian Oscillation are strongly coupled with convective phenomena [59, 101, 112].

The outline of this chapter is the following. First, we describe in Sect. 3.2 the most recent theoretical results regarding the internal variability of the mid-latitude wind-driven circulation, viewed as a problem in nonlinear fluid mechanics. These results rely to a large extent on autonomous DDS theory [4, 69]. Next, we summarize in Sect. 3.3 the key concepts and methods of NDS [137] and RDS [5] theory.

With the results of Sects. 3.2 and 3.3 in hand, we address in Sect. 3.4 the changes in this purely oceanic variability as a result of time-dependent forcing, on the one hand, and truly coupled ocean–atmosphere variability, on the other. Much of the material in this section is new [50, 125, 168]. A summary and an outlook on future work follow in Sect. 3.5. Rigorous mathematical results appear in Appendices 1–3, which follow closely [64] (Appendix 1) and [23] (Appendices 2 and 3).

3.2   Internal Variability of the Wind-Driven Ocean Circulation

3.2.1   Observations

To a first approximation, the main near-surface currents in the oceans are driven
by the mean effect of the winds. The trade winds near the equator blow mainly
from east to west and are also called the tropical easterlies. In mid-latitudes, the
dominant winds are the prevailing westerlies, and towards the poles the winds are
easterly again. Thus an idealized version of wind effects on the ocean must involve
zonally oriented winds that blow westward at low and high latitudes, and eastward
in mid-latitudes.

Five of the strongest near-surface, mid-and-high-latitude currents are the Antarctic Circumpolar Current, the Gulf Stream and the Labrador Current in the North
Atlantic, and the Kuroshio and Oyashio off Japan. These currents are all clearly
visible in Fig. 14.

The Gulf Stream [154] is an oceanic jet with a strong influence on the climate of
eastern North America and of western Europe. From Mexico's Yucatan Peninsula,
the Gulf Stream flows north through the Florida Straits and along the East Coast of
the United States. Near Cape Hatteras, it detaches from the coast and begins to drift
off into the North Atlantic towards the Grand Banks near Newfoundland. Actually,
the Gulf Stream is part of a larger, gyre-like current system, which includes the
North Atlantic Drift, the Canary Current and the North Equatorial Current. It is also
coupled with the pole-to-pole overturning circulation.

The Coriolis force is responsible for the so-called Ekman transport, which
deflects water masses orthogonally to the near-surface wind direction and to the right
[56, 67, 120]. In the North Atlantic, this Ekman transport creates a divergence and a
convergence of near-surface water masses, respectively, resulting in the formation of
two oceanic gyres: a smaller, cyclonic one in subpolar latitudes, the other larger and
anticyclonic in the subtropics. This type of *double-gyre* circulation characterizes all
mid-latitude ocean basins, including the South Atlantic, as well as the North and
South Pacific.

The double-gyre circulation is intensified as the currents approach the East Coast
of North America due to the β-effect. This effect arises primarily from the variation
of the Coriolis force with latitude, while the oceans' bottom topography also contributes to it. The former, planetary β-effect is of crucial importance in geophysical
flows and induces free Rossby waves propagating westward [56, 67, 120].

The currents along the western shores of the North Atlantic and of the other
mid-latitude ocean basins exhibit boundary-layer characteristics and are commonly
called western boundary currents. The northward-flowing Gulf Stream and the
southward-flowing Labrador Current extension meet near Cape Hatteras and give
rise to a strong, mainly eastward jet that eventually reaches Scandinavia. In all
four mid-latitude ocean basins—i.e., the North and South Atlantic, as well as the
North and South Pacific—the confluence of a warm, poleward current with a cold,
equatorward one produces an intense eastward jet. The formation of this jet and of

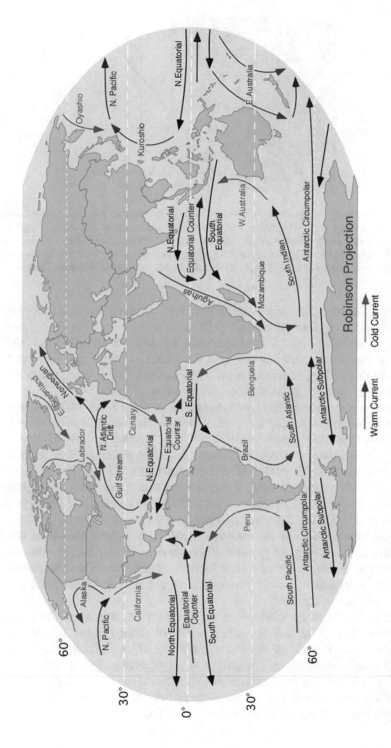

Fig. 14 A map of the main oceanic currents: warm currents in red and cold ones in blue. Reproduced from [64], with permission from Elsevier (Color figure online)

Fig. 15 A satellite image of the sea surface temperature (SST) field over the northwestern North Atlantic (in false color, from the U.S. National Oceanic and Atmospheric Administration), together with a sketch of the associated double-gyre circulation (white arrows). A highly simplified and smoothed view of the amount of potential vorticity injected into the ocean circulation by the wind stress is shown in the sketch to the right. Reproduced from [64], with permission from Elsevier

the associated recirculation vortices near the western boundary, to either side of the jet, is mostly driven by internal, nonlinear effects.

Figure 15 illustrates how these large-scale wind-driven oceanic flows self-organize, as well as the resulting eastward jet. Different spatial and time scales contribute to this self-organization: so-called mesoscales eddies in the ocean play the role of the synoptic-scale weather systems in the atmosphere.

Two cold rings and a warm one are clearly visible in Fig. 15. Such warm and cold eddies last for several months up to a year and have a diameter of about 100 km vs. the much larger horizontal dimensions of atmospheric weather systems, of the order of 1000 km. In fact, the scale of roughly 100 km that characterizes these eddies in the ocean is given by the Rossby radius of deformation that was introduced in Sect. 2. This scale is much smaller than that of atmospheric weather systems, although the basic physics is the same. The difference in scale of the oceanic and atmospheric Rossby radius is due to the different stratification of the two fluid media.

The characteristic scale of the eastward oceanic jet that detaches from Cape Hatteras and of the two opposite gyres on either side of it—sketched as the recurving white arrows in the figure—is of several thousand kilometers. These larger-scale oceanic features exhibit their own intrinsic dynamics on time scales of several years to possibly one or two decades.

A striking feature of the wind-driven circulation is the existence of two well-known North Atlantic Oscillations, with a period of about 7–8 and 14–15 years, respectively. Data analysis of various climatic variables, such as sea surface temperature (SST) over the North Atlantic or sea level pressure (SLP) over western Europe [36, 110, 171] and local surface air temperatures in Central England [126], as well as of proxy records, such as tree rings in Britain, travertine concretions in southeastern France [44], and Nile floods over the last millennium or so [85], all exhibit strikingly robust oscillatory behavior with a 7–8-year period and, to a lesser extent, with a 14-year period. Variations in the path and intensity of the Gulf Stream are most likely to exert a major influence on the climate in this part of the world [154]. These climatic implications add to the intrinsic interest of theoretical studies of the low-frequency variability of the oceans' double-gyre circulation.

Given the complexity of the processes involved, climate studies have been most successful when using not just a single model but a full hierarchy of models, from the simplest to models to the most detailed GCMs [53, 59]. In the following, we describe one of the simplest models of the hierarchy used in studying this problem, which still retains many of its essential features.

3.2.2 A Simple Model of the Double-Gyre Circulation

The simplest model that includes many of the mechanisms described above is governed by the barotropic *quasi-geostrophic* (QG) equations. The term geostrophic refers to the fact that large-scale rotating flows tend to run parallel to, rather than perpendicular to constant-pressure contours; in the oceans, these contours are associated with the deviation from rest of the surfaces of equal water mass, due to Ekman pumping. Geostrophic balance implies in particular that the flow is divergence-free. The term barotropic, as opposed to baroclinic, has a slightly different meaning in geophysical fluid dynamics than in engineering fluid mechanics: it means that the model describes a single fluid layer of constant density and therefore the solutions do not depend on depth [56, 67, 120].

We consider an idealized, rectangular basin geometry and simplified forcing that mimics the distribution of vorticity due to the wind stress, as sketched to the right of Fig. 15. In our idealized model, the amounts of subpolar and subtropical vorticity injected into the basin are equal and the rectangular domain $\Omega = (0, L_x) \times (0, L_y)$ is symmetric about the axis of zero wind stress curl.

The barotropic two-dimensional (2-D) QG equations in this idealized setting are:

$$q_t + J(\psi, q) - \nu \Delta^2 \psi + \mu \Delta \psi = -\tau \sin \frac{2\pi y}{L_y},$$
$$q = \Delta \psi - \lambda_R^{-2} \psi + \beta y. \tag{101}$$

Here q and ψ are the potential vorticity and the streamfunction, respectively, and the Jacobian J gives the advection of potential vorticity by the flow, $J(\psi, q) = \psi_x q_y - \psi_y q_x = \mathbf{u} \cdot \nabla q$, where $\mathbf{u} = (-\psi_y, \psi_x)$, x points east and y points north.

The physical parameters are the strength of the planetary vorticity gradient β, the Rossby radius of deformation λ_R^{-2}, the eddy-viscosity coefficient ν, the bottom friction coefficient μ, and the wind-stress intensity τ. We use here free-slip boundary conditions $\psi = \Delta^2 \psi = 0$; the qualitative results described below do not depend on the particular choice of homogeneous boundary conditions [39, 76].

We consider the nonlinear PDE system (101) as an infinite-dimensional dynamical system and study its bifurcations as the parameters change. Two key parameters are the wind stress intensity τ and the eddy viscosity ν. An important property of (101) is its mirror symmetry in the $y = L_y/2$ axis. This symmetry can be expressed as invariance with respect to the discrete $\mathbb{Z}_2$ group $\mathcal{S}$, given by

$$\mathcal{S}[\psi(x, y)] = -\psi(x, L_y - y); \tag{102}$$

any solution of (101) is thus accompanied by its mirror-conjugated solution. Hence the prevailing bifurcations are of either the symmetry-breaking or the Hopf type.

3.2.3   Bifurcations in the Double-Gyre Problem

The development of a comprehensive nonlinear theory of the double-gyre circulation over the last two decades has gone through four main steps. These four steps can be followed through the bifurcation tree in Fig. 16.

Symmetry-Breaking Bifurcation

The "trunk" of the bifurcation tree is plotted as the solid blue line in the lower part of the figure. When the forcing τ is weak or the dissipation ν is large, there is only one steady solution, which is antisymmetric with respect to the mid-axis of the basin. This solution exhibits two large gyres, along with their β-induced western boundary currents. Away from the western boundary, such a near-linear solution (not shown) is dominated by so-called Sverdrup balance between wind stress curl and the meridional mass transport [67, 156].

The first generic bifurcation of this QG model was found to be a genuine pitchfork bifurcation that breaks the system's symmetry as the nonlinearity becomes large enough with increasing wind stress intensity τ [17, 75, 76]. As the wind stress increases, the near-linear Sverdrup solution that lies along the solid blue line in the figure develops an eastward jet along the mid-axis, which penetrates farther into the domain and also forms two intense recirculation vortices, on either side of the jet and near the western boundary of the domain.

The resulting more intense, and hence more nonlinear solution is still antisymmetric about the mid-axis, but loses its stability for some critical value of the wind-stress intensity, $\tau = \tau_P$. This value is indicated by the filled square on the symmetry axis of Fig. 16 and is labeled "**Pitchfork**" in the figure.

A pair of mirror-symmetric solutions emerges and it is plotted as the two red solid lines in the figure's lower part. The streamfunction fields associated with the two stable steady-state branches have a rather different vorticity distribution and they are plotted in the two small panels to the upper-left and upper-right of Fig. 16. In particular, the jet in such a solution exhibits a large meander, reminiscent of the one seen in Fig. 15 just downstream of Cape Hatteras; note that the colors in Fig. 16 were chosen to facilitate the comparison with Fig. 15. These asymmetric flows are characterized by one recirculation vortex being stronger in intensity than

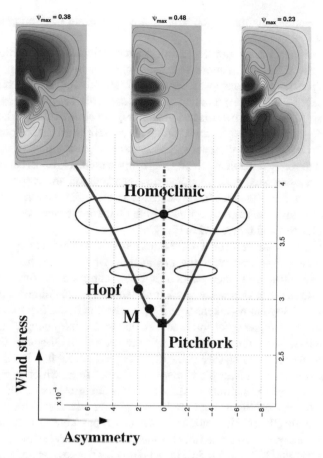

Fig. 16 Generic bifurcation diagram for the barotropic QG model of the double-gyre problem: the asymmetry of the solution is plotted versus the intensity of the wind stress τ. The streamfunction field is plotted for a steady-state solution associated with each of the three branches; positive values in red and negative ones in blue. After [143]

the other, which deflects the jet accordingly either to the southeast, as is the case in the observations for the North Atlantic, or to the northeast.

Gyre Modes

The next step in the theoretical treatment of the problem was taken in part concurrently with the first one above [75, 76] and in part shortly thereafter [40, 138, 150]. It involved the study of time-periodic instabilities through Hopf bifurcation from either an antisymmetric or an asymmetric steady flow. Some of these studies treated wind-driven circulation models limited to a stand-alone, single gyre [121, 138]; such a model concentrates on the larger, subtropical gyre while neglecting the smaller, subpolar one.

The overall idea was to develop a full, generic picture of the time-dependent behavior of the solutions in more turbulent regimes, by classifying the various instabilities in a comprehensive way. However, it quickly appeared that a particular kind of instability leads to so-called *gyre modes* [76, 150], and was prevalent across the full hierarchy of models of the double-gyre circulation; furthermore, this instability triggers the lowest nonzero frequency present in all such models [37, 39].

These gyre modes always appear after the first pitchfork bifurcation, and it took several years to really understand their genesis: gyre modes arise as two eigenvalues merge—one is associated with a symmetric eigenfunction and responsible for the pitchfork bifurcation, the other is associated with an antisymmetric eigenfunction [139]; this merging is marked by a filled circle on the left branch of antisymmetric stationary solutions and labeled **M** in Fig. 16.

Such a phenomenon is not a bifurcation *stricto sensu*: one has topological C^0-equivalence before and after the eigenvalue merging, but not from the C^1 point of view. Still, this phenomenon is quite common in small-dimensional dynamical systems with symmetry, as exemplified by the unfolding of codimension-2 bifurcations of Bogdanov-Takens type [69]. In particular, the fact that gyre modes trigger the lowest-frequency of the model is due to the frequency of these modes growing quadratically from zero until nonlinear saturation. Of course these modes, in turn, become unstable shortly after the merging, through a Hopf bifurcation indicated in Fig. 16 by a heavy dot marked "**Hopf**," from which a stylized limit cycle emerges. A mirror-symmetric **M** and **Hopf** bifurcation also occur on the right branch of stationary solutions, but have been omitted in the figure for visual clarity.

More generally, Hopf bifurcations of various types give rise to features that recur more-or-less periodically in fully turbulent planetary-scale flows, atmospheric, oceanic and coupled [37, 39, 54, 56]. In the climate sciences, one commonly refers to this type of near-periodic recurrence as *low-frequency variability (LFV)*.

Global Bifurcations

The importance of the gyre modes was further confirmed through an even more puzzling discovery. Several authors realized, independently of each other, that the low-frequency dynamics of their respective double-gyre models was driven by intense relaxation oscillations of the jet [19, 106, 111, 140–143]. These relaxation oscillations, already described in [76, 150], were now attributed to a homoclinic bifurcation, with a global character in phase space [56, 69]. In effect, the QG model reviewed here undergoes a genuine homoclinic bifurcation that is generic across the full hierarchy of double-gyre models.

This bifurcation is due to the growth and eventual merging of the limit cycles that arise from the two mutually symmetric Hopf bifurcations. It is marked in the figure by a filled circle and a stylized lemniscate, and labeled "**Homoclinic**." This global bifurcation is associated with chaotic behavior of the flow due to the Shilnikov phenomenon [111, 143], which induces horseshoes in phase space.

The connection between such homoclinic bifurcations and gyre modes was not immediately obvious, but Simonnet et al. [143] emphasized that the two were part of a single, global dynamical phenomenon. The homoclinic bifurcation indeed results from the unfolding of the gyre modes' limit cycles. This familiar dynamical scenario is again well illustrated by the unfolding of a codimension-2 Bogdanov-Takens bifurcation, where the homoclinic orbits emerge naturally.

Since homoclinic orbits have an infinite period, it was natural to hypothesize that the gyre-mode mechanism, in this broader, global-bifurcation context, gave rise to the observed 7- and 14-year North Atlantic oscillations. Although this hypothesis may appear a little farfetched—given the simplicity of the double-gyre models analyzed so far—it is reinforced by results with much more detailed model in the hierarchy, cf. [37, 39] and Sect. 3.4 herein.

The successive-bifurcation theory appears therewith to be fairly complete for barotropic, single-layer models of the double-gyre circulation. This theory also provides a self-consistent, plausible explanation for the climatically important 7- and 14-year oscillations of the oceanic circulation and the related atmospheric phenomena in and around the North Atlantic basin [36, 37, 39, 47–50, 85, 110, 126, 142, 143, 171]. The dominant 7- and 14-year modes of this theory survive, moreover, perturbation by seasonal-cycle changes in the intensity and meridional position of the westerly winds [155].

In baroclinic models, with two or more active layers of different density, baroclinic instabilities [8, 39, 48, 56, 67, 88, 120, 121, 142, 154] surely play a fundamental role, as they do in the observed dynamics of the oceans. However, it is not known to what extent baroclinic instabilities can destroy gyre-mode dynamics. The difficulty lies in a deeper understanding of the so-called rectification process [81], which arises from the nonzero mean effect of the baroclinic component of the flow.

Roughly speaking, rectification drives the dynamics far away from any stationary solutions. In this situation, dynamical systems theory by itself cannot be used as a full explanation of complex, observed behavior resulting from successive bifurcations that are rooted in simple stationary or periodic solutions. Other tools from statistical mechanics and nonequilibrium thermodynamics should, therefore, be considered [13, 46, 98, 99, 102, 131, 160]. Combining these tools with those of the successive-bifurcation approach may eventually lead to a more general and complete physical characterization of gyre modes in realistic models.

3.3   Non-autonomous and Random Dynamical Systems

An additional way of improving upon the simple results so far is to include time-dependent forcing and stochasticity. As discussed in Sect. 3.1, the appropriate general framework for doing this is the one provided by NDS and RDS theory

[10, 23, 43, 64]. In the present section, we summarize this mathematical framework in as straightforward a way as possible, and start with some simple ideas about deterministic vs. stochastic modeling.

3.3.1 Background and Motivation

We are interested in behavior that is robust across a full hierarchy of climate models in general, and of oceanic double-gyre models in particular. Moreover, we would like this robust behavior of the models to be recognizable in the very large and ever-increasing mass of observational data. Finally, the features in whose robustness we are most interested are those that obtain over long time intervals.

Classical DDS theory, going back to H. Poincaré [127], is essentially a geometric approach to studying the asymptotic, long-term properties of solutions to autonomous ODE systems in phase space. Extensions to nonlinear PDE systems are covered, for instance, in [158]. To apply the theory in a reliable manner to a set of complex physical phenomena, one needs a criterion for the robustness of a given model within a class of dynamical systems. Such a criterion should help one deal with the inescapable uncertainties in model formulation, whether due to incomplete knowledge of the governing laws or inaccuracies in determining model parameters.

In this context, A. A. Andronov and L. S. Pontryagin [2] introduced the concept of *structural stability* for classifying dynamical systems. Structural stability means that a small, continuous perturbation of a given system preserves its dynamics up to a *homeomorphism*, i.e., up to a one-to-one continuous change of variables that transforms the phase portrait of our system into that of a nearby system; thus fixed points go into fixed points, limit cycles into limit cycles, etc. Closely related is the notion of *hyperbolicity* formulated by S. Smale [147]. A system is hyperbolic if, (very) loosely speaking, its limit set can be continuously decomposed into invariant sets that are either contracting or expanding; see [80] for more rigorous definitions.

A very simple example is the phase portrait in the neighborhood of a fixed point of saddle type. The Hartman-Grobman theorem states that the dynamics in this neighborhood is structurally stable. The converse statement, i.e. whether structural stability implies hyperbolicity, is still an open question; the equivalence between structural stability and hyperbolicity has only been shown in the C^1 case, under certain technical conditions [103, 117, 130, 132]. Bifurcation theory is quite complete and satisfying in the setting of hyperbolic dynamics. Problems with hyperbolicity and bifurcations arise, however, when one deals with more complicated limit sets.

Hyperbolicity was introduced initially to help pursue the "dynamicist's dream" of finding—in an abstract space $\mathcal{X}$ of all possible dynamical systems with given dimension and regularity—an open and dense set S consisting of structurally stable ones. For this set to be open and dense means, roughly speaking, that any dynamical system in $\mathcal{X}$ can be approximated by structurally stable ones in S, while systems in the complement of S are negligible in a suitable sense.

Smale conjectured that hyperbolic systems form an open and dense set in the space of all C^1 dynamical systems. If this conjecture were true then hyperbolicity would be typical of all dynamics. Unfortunately, though, this conjecture is only true for one-dimensional dynamics and flows on disks and surfaces [122]. Smale [146] himself found several counterexamples to his conjecture. Newhouse [113] was able to generate open sets of nonhyperbolic diffeomorphisms using homoclinic tangencies. For the physicist, it is even more striking that the famous Lorenz attractor [96] is structurally unstable. Families of Lorenz attractors, classified by topological type, are not even countable [68, 170]. In each of these examples, we observe chaotic behavior in a nonhyperbolic situation, i.e., *nonhyperbolic chaos.*

Nonhyperbolic chaos appears, therefore, to be a severe obstacle to any easy classification of dynamic behavior. As mentioned by J. Palis [117], A. N. Kolmogorov already suggested at the end of the sixties that "the global study of dynamical systems could not go very far without the use of new additional mathematical tools, like probabilistic ones."

Once more, Kolmogorov showed prophetic insight, and nowadays the concept of *stochastic stability* is an important tool in the study of genericity and robustness for dynamical systems. A system is stochastically stable if its Sinai-Ruelle-Bowen (SRB) measure [145] is stable with respect to stochastic perturbations, and the SRB measure is given by $\lim_{n \to \infty} \frac{1}{n} \sum_i \delta_{z_i}$, with z_i being the successive iterates of the dynamics. This measure is obtained intuitively by allowing the entire phase space to flow onto the attractor [45].

To replace the failed program of classifying dynamical systems based on structural stability and hyperbolicity, J. Palis [117] formulated his so-called *global conjecture*: The set S of systems satisfying the following conditions is dense in the C^r-topology for $r \geq 1$. The three conditions are:

(1) the system has only finitely many attractors, i.e. periodic or chaotic sinks;
(2) the union of the corresponding basins of attraction has full Lebesgue measure in $\mathcal{X}$; and
(3) each attractor is stochastically stable in its basin of attraction.

Stochastic stability is thus based on ergodic theory. We would like to consider a more geometric approach, which can provide a coarser, more robust classification of climate models across a full hierarchy of such models. In this section, we propose such an approach, based on RDS theory [5, 26, 33, 91, and references therein].

RDS theory describes the behavior of dynamical systems subject to external stochastic forcing; its tools have been developed to help study the geometric properties of stochastic differential equations (SDEs). RDS theory is thus the stochastic, SDE counterpart of the geometric theory of ODEs, as presented for instance in [4]. This approach provides a rigorous mathematical framework for a stochastic form of robustness, while the more traditional, topological concepts do not seem to be appropriate for the climate sciences, given the prevalence of nonhyperbolic chaos.

3.3.2 Pullback and Random Attractors

Stochastic parametrizations for GCMs aim at compensating for our lack of detailed knowledge on small spatial scales in the best way possible [92–94, 118, 119, 152]. The underlying assumption is that the associated time scales are also much shorter than the scales of interest and, therefore, the lag correlation of the phenomena being parametrized is negligibly small. Such assumptions clearly hold, for instance, for the spatial and temporal scales of cloud processes, whose parametrization poses a notorious obstacle to realistic climate simulation and prediction [71–74]. Stochastic parametrizations thus essentially transform a deterministic autonomous system into a nonautonomous one, subject to random forcing.

Explicit time dependence, whether deterministic or stochastic, in a dynamical system immediately raises a technical difficulty. Indeed, the classical notion of attractor is not always relevant, since any object in phase space is moving with time and the concept of forward asymptotics, as considered in autonomous DDS theory, is meaningless. One needs therefore another notion of attractor. In the deterministic nonautonomous framework, the appropriate notion is that of a *pullback attractor* [16, 83, 129], which we present below. The closely related notion of *random attractor* in the stochastic framework is also explained briefly below, with further details given in Appendix 1.

Pullback Attraction

Before defining the notion of pullback attractor, let us recall some basic facts about nonautonomous dynamical systems. Consider the ODE

$$\dot{\mathbf{x}} = \mathbf{f}(\mathbf{x}, t) \tag{103}$$

on a vector space X; this space could even be infinite-dimensional, if we were dealing with partial or functional differential equations, as is often the case in fluid-flow and climate problems. Rigorously speaking, we cannot associate a dynamical system acting on X with a nonautonomous ODE; nevertheless, in the case of unique solvability of the initial-value problem, we can introduce a two-parameter family of operators $\{S(t, s)\}_{t \geq s}$ acting on X, with s and t real, such that $S(t, s)x(s) = x(t)$ for $t \geq s$, where $x(t)$ is the solution of the Cauchy problem with initial data $x(s)$. This family of operators satisfies $S(s, s) = \mathrm{Id}_X$ and $S(t, \tau) \circ S(\tau, s) = S(t, s)$ for all $t \geq \tau \geq s$, and all real s. Such a family of operators is called a "process" by Sell [137]. It extends the classical notion of the resolvent of a nonautonomous linear ODE to the nonlinear setting.

We can now define the pullback attractor as the family of invariant sets $\{\mathcal{A}(t) : -\infty < t < +\infty\}$ that satisfy, for every real t and all $\mathbf{x_0}$ in X:

$$\lim_{s \to -\infty} \text{dist}\,(S(t, s)\mathbf{x_0}, \mathcal{A}(t)) = 0. \tag{104}$$

"Pullback" attraction does not involve running time backwards; it corresponds instead to the idea of measurements being performed at present time t in an experiment that was started at some time $s < t$ in the past: the experiment has been running for long enough, and we are thus looking at the present moment for an "attracting state." Note that there exist several ways of defining a pullback attractor—the one retained here is a local one, cf. [89, and references therein]; see [7] for further information on nonautonomous dynamical systems in general.

A simple example will be helpful for the general reader. Consider the scalar linear ODE

$$\dot{x} = -\alpha x + \sigma t, \tag{105}$$

with both α and σ positive. Here $\alpha > 0$ implies the system is *dissipative,* a crucial property both mathematically and physically. In the climate context, its importance has been emphasized by Lorenz [96], Ghil and Childress [56], among others.

The situation is depicted in Fig. 17. The PBA is easily computed in this case, and it is given by

$$\mathcal{A}(t) \equiv a(t) = \frac{\sigma}{\alpha}(t - \frac{1}{\alpha}). \tag{106}$$

The figure clearly shows that the approximation of the PBA at a given time $t = t_1$ or $t = t_2$ improves as we "pull back" further, from $s = s_1$ to $s = s_2$ (red vs. blue orbits in the figure); it also shows that the accuracy of the approximation depends essentially on the time interval $|t - s|$, rather than on t and s separately. In practice, the characteristic time for obtaining an approximation with given accuracy will depend largely on the rate of dissipation α.

In the stochastic context, noise forcing is modeled by a stationary stochastic process. If the deterministic dynamical system of interest is coupled to this stochastic process in a reasonable way—defined below by the "cocycle property"— then random pullback attractors may appear. These pullback attractors will exist for almost each sample path of the driving stochastic process, so that the same probability distribution governs both sample paths and their corresponding pullback attractors. A more detailed and rigorous explanation is given in Appendix 1.

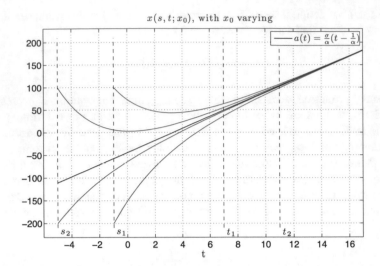

Fig. 17 Pullback attractor (PBA) for a scalar linear equation, given by the solid black line. We wish to observe the PBA at times $t = t_1, t_2$, marked by dashed vertical lines, and consider the convergence to the straight line of orbits started at past times $s = s_1, s_2$; sample orbits for s_1 and s_2 are plotted in red and blue, respectively. Courtesy of M. D. Chekroun (Color figure online)

Roughly speaking, the concept of a *random attractor* provides a geometric framework for the description of asymptotic regimes in the context of stochastic dynamics. Figure 17 showed how additive and linear deterministic forcing modifies a fixed point. We illustrate next, following [23, 54], the way that multiplicative stochastic forcing modifies a strange attractor.

The Stochastically Perturbed Lorenz Model

We consider in this section the Lorenz convection model [96] and its randomly forced version studied in [23]. The model is governed by the well-known three nonlinear, coupled ODEs that are now forced by linearly multiplicative white noise [5, 26].

$$[\mathbf{SLM}] \quad \begin{cases} dx = s(y - x)dt + \sigma x \, dW_t, \\ dy = (rx - y - xz)dt + \sigma y \, dW_t, \\ dz = (-bz + xy)dt + \sigma z \, dW_t. \end{cases} \tag{107}$$

In the deterministic context, purely geometric map models were proposed in the 1970s [69] to interpret the dynamics observed numerically by E. N. Lorenz in [96].

These geometric models attracted considerable attention and it was shown that they possess a unique SRB measure [28, 82], i.e., a time-independent measure that is invariant under the flow and has conditional measures on unstable manifolds that are absolutely continuous with respect to Lebesgue measure [45]. This result has been extended recently to the Lorenz flow [96] itself, in which the SRB measure is supported by a strange attractor of vanishing volume [3, 161].

Even though this result was only proven recently, the existence of such an SRB measure was suspected for a long time and has motivated several numerical studies to compute a probability density function (PDF) associated with the Lorenz model, by filtering out the stable manifolds [42, 109, and references therein]. The Lorenz attractor is then approximated by a two-dimensional manifold, called the *branched manifold* [69], which supports this PDF. Based on such a strategy, Dorfle and Graham [42] showed that the stationary solution of the Fokker-Planck equation for the Lorenz model perturbed by additive white noise possesses a density with two components: the PDF of the deterministic system supported by the branched manifold plus a narrow Gaussian distribution transversal to that manifold.

It follows that, in the presence of additive noise, the resulting PDF looks very much like that of the unperturbed system, only slightly fuzzier: the noise smoothes the small-scale structures of the attractor. More generally, this smoothing appears in the classical, *forward approach* that only considers forward asymptotics, with $t \rightarrow +\infty$—as opposed to the *pullback approach*, introduced here in the previous Paragraph, in which one considers also the pullback asymptotics of $s \rightarrow -\infty$— and it does so for a broad class of additive as well as multiplicative noises, in the sense of [5]. More precisely, this smoothing occurs provided that the diffusion terms due to the stochastic components in the Fokker-Planck equation associated with the SDE system under study are sufficiently non-degenerate; see [23, Appendix C and references therein].

Hörmander's theorem guarantees that this is indeed the case for hypoelliptic SDEs [6]. The corresponding non-degeneracy conditions allow one to regularize the stationary solutions of the *transport equation*

$$\partial_t p(\mathbf{x}, t) = -\nabla \cdot (p(\mathbf{x}, t) \mathbf{F}(\mathbf{x})), \tag{108}$$

which is the counterpart of the Fokker-Planck equation in the absence of noise; a measure-theoretic justification for it can be found, for instance in [90, p. 210].

This transport equation is also known as the Liouville equation and it provides the probability density at time t of $S(t)x$ when the initial state x is sampled from a probability measure that is absolutely continuous with respect to Lebesgue measure; here $\{S(t)\}_{t \in \mathbb{R}}$ is the flow of $\dot{\mathbf{x}} = \mathbf{f}(\mathbf{x})$, for some sufficiently smooth vector field $\mathbf{f}$ on $\mathbb{R}^d$. As a matter of fact, when $\mathbf{f}$ is dissipative and the dynamics associated with it is chaotic, the stationary solutions of (108) are very often singular with respect to Lebesgue measure; these solutions are therefore expected to be SRB measures. For a broad class of noises—such as those that obey a hypoellipticity

(a)   (b)

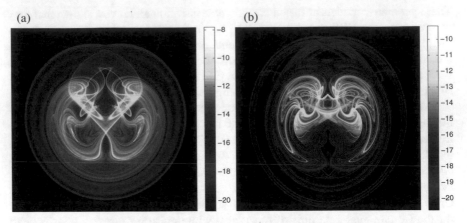

Fig. 18 Snapshot of the Lorenz [96] model's random attractor $\mathcal{A}(\omega)$ and of the corresponding sample measure μ_ω, for a given, fixed realization ω. The figure corresponds to projection onto the (y, z) plane, i.e. $\int \mu_\omega(x, y, z)\mathrm{d}x$. One billion initial points have been used in both panels and the pullback attractor is computed for $t = 40$. The parameter values are the classical ones—$r = 28$, $s = 10$, and $b = 8/3$, while the time step is $\Delta t = 5 \cdot 10^{-3}$. The color bar to the right of each panel is on a log-scale and quantifies the probability to end up in a particular region of phase space. Both panels use the same noise realization ω but with noise intensity (**a**) $\sigma = 0.3$ and (**b**) $\sigma = 0.5$. Notice the interlaced filamentary structures between highly (yellow) and moderately (red) populated regions; these structures are much more complex in panel (**b**), where the noise is stronger. Weakly populated regions cover an important part of the random attractor and are, in turn, entangled with (almost) zero-probability regions (black). After [23] (Color figure online)

condition—the forward approach leads us to suspect that noise effects tend to remove the singular aspects with respect to Lebesgue measure. This smoothing aspect of random perturbations is often useful in the theoretical understanding of any stochastic system, in particular in the analysis of the low- and higher-order moments, which have been thoroughly studied in various contexts.

For chaotic systems subject to noise, however, this noise-induced smoothing observed in the forward approach compresses a lot of crucial information about the dynamics itself; quite to the contrary, the pullback approach brings this information into sharp focus. A quick look at Figs. 18a, b and 19 is already enlightening in this respect. All three figures refer to the invariant measure μ_ω supported by the random attractor of our stochastic Lorenz model **[SLM]**. This model obeys the following three SDEs: In system (107), each of the three equations of the classical, deterministic model [96] is perturbed by linearly multiplicative noise in the Itô sense, with W_t a Wiener process and $\sigma > 0$ the noise intensity. The other parameter values are the standard ones for chaotic behavior [56], and are given in the caption of Fig. 18.

Figure 18a, b show two snapshots of the sample measure μ_ω supported by the random attractor of **[SLM]**—for the same realization ω but for two different noise intensities, $\sigma = 0.3$ and 0.5, while Fig. 19 provides four successive snapshots of

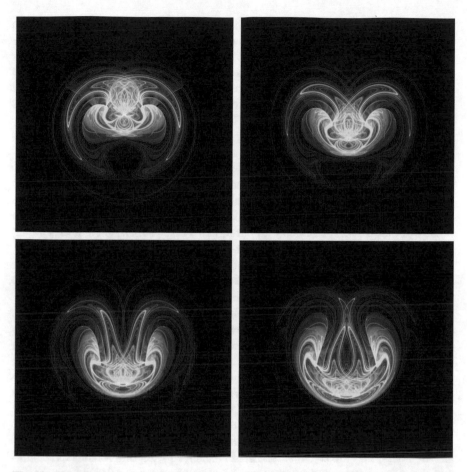

Fig. 19 Four snapshots of the random attractor and sample measure supported on it, for the same parameter values as in Fig. 18. The time interval Δt between two successive snapshots—moving from left to right and top to bottom—is $\Delta t = 0.0875$. Note that the support of the sample measure may change quite abruptly, from time to time, cf. short video in [23, Supplementary Material] for details. Reproduced from [23], with permission from Elsevier

$\mu_{\theta_t \omega}$, for the same noise intensity $\sigma = 0.5$ as in Fig. 18b, but with $t = t_0 + k \Delta t$ and $k = 0, 1, 2, 3$ for some t_0.

The sample measures in these three figures, and in the associated short video given in the SM, exhibit amazing complexity, with fine, very intense filamentation; note logarithmic scale on color bars in the three figures. There is no fuzziness whatsoever in the topological structure of this filamentation, which evokes the Cantor-set foliation of the deterministic attractor [69]. Such a fine structure strongly suggests that these measures are supported by an object of vanishing volume.

Much more can be said, in fact, about these objects. RDS theory offers a rigorous way to define random versions of stable and unstable manifolds, via the Lyapunov spectrum, the Oseledec multiplicative theorem, and a random version of the Hartman-Grobman theorem [5]. These random invariant manifolds can support measures, like in the deterministic context. When the sample measures μ_ω of an RDS have absolutely continuous conditional measures on the random unstable manifolds, then μ_ω is called a *random SRB measure*.

One can prove rigorously, by relying on Theorem B of [91], that the sample measures of the discretized stochastic system obtained from the **[SLM]** model share the SRB property. Indeed, it can be shown that a *Hörmander hypoellipticity condition* is satisfied for our discretized **[SLM]** model, thus ensuring that the random process generated by this model has a smooth density $p(t, \mathbf{x})$ [84]; see [23, Appendix C1] for the details. Standard arguments [149] can then be used to prove that the stationary solution ρ of our model's Fokker-Planck equation is in fact absolutely continuous with respect to Lebesgue measure.

Since these simulations exhibit exactly one positive Lyapunov exponent, the absolute continuity of ρ implies that the sample measures seen in Figs. 18 and 19 are, actually, good numerical approximations of a genuine random SRB measure for our discretized **[SLM]**, whenever δt is sufficiently small; see also the next section. In fact, Ledrappier and Young's [91] Theorem B is a powerful result, which clearly shows that—in noisy systems, and subject to fairly general conditions—chaos can lead to invariant sample measures with the SRB property; see also [23, Appendix C2]. It is striking that the same noise-induced smoothing that was "hiding" the dynamics in the forward approach allows one here to exhibit the existence of an SRB measure from a pullback point of view, and thus to compute explicitly the unstable manifold supporting this invariant measure.

Note that, since the sample measures associated with the discrete **[SLM]** system are SRB here, they are so-called *physical measures* [5, 23, and references therein] and can thus be computed at any time t by simply flowing a large set of initial data from the remote past $s \ll t$ up till t, for a fixed realization ω; this is exactly how Figs. 18 and 19 were obtained. Given the SRB property, the nonzero density supported on the model's unstable manifolds delineates numerically these manifolds; Figs. 18 and 19 provide therefore an approximation of the *global* random attractor of our stochastic Lorenz system.

Finally, these random measures are Markovian, in the sense that they are measurable with respect to the past σ-algebra of the noise [5]. The latter statement results directly from the fact that these measures are physical, and thus satisfy the required measurability conditions in the pullback limit. The information about the moments that is available in the classical Fokker-Planck approach is complemented here by information about the *pathwise moments*. These pathwise statistics are naturally associated with the sample measures—when the latter are SRB—by taking appropriate averages.

The evolution of the sample measures $\mu_{\theta_t \omega}$—as apparent from the short video in [23, Supplementary Material]—is quite complex, and two types of motion are present. First, a pervasive "jiggling" of the overall structure can be traced back to the roughness of the Wiener process W_t and to the multiplicative way it enters into the [SLM] model. Second, there is a smooth, regular low-frequency motion present in the evolution of the sample measures, which seems to be driven by the deterministic system's unstable limit cycles and is thus related to the well-known lobe dynamics. The latter motion is clearly illustrated in Fig. 19.

More generally, it is worth noting that this type of low-frequency motion seems to occur quite often in the evolution of the samples measures of chaotic systems perturbed by noise; it appears to be related to the recurrence properties of the unperturbed deterministic flow, especially when energetic oscillatory modes characterize the latter. To the best of our knowledge, there are no rigorous results on this type of phenomenon in RDS theory.

Besides this low-frequency motion, abrupt changes in the global structure occur from time to time, with the support of the sample measure either shrinking or expanding suddenly. These abrupt changes recur frequently in the video associated with Fig. 19, which reproduces a relatively short sequence out of a very long stochastic model integration; see SM.

As the noise intensity σ tends to zero, the sample-measure evolution slows down, and one recovers numerically the measure of the deterministic Lorenz system (not shown). This convergence as $\sigma \to 0$ may be related to the concept of *stochastic stability* [82, 172]. Such a continuity property of the sample measures in the zero-noise limit does not, however, hold in general; it depends on properties of the noise, as well as of the unperturbed attractor [15, 30, 34].

As stated in the theoretical section, the forward approach is recovered by taking the expectation, $\mathbb{E}[\mu_\bullet] := \int_\Omega \mu_\omega \mathbb{P}(d\omega)$, of these invariant sample measures. In practice, $\mathbb{E}[\mu_\bullet]$ is closely related to ensemble or time averages that typically yield the previously mentioned PDFs. In addition, when the random invariant measures are Markovian and the Fokker-Planck equation possesses stationary solutions, $\mathbb{E}[\mu_\bullet] = \rho$, where ρ is such a solution. Subject to these conditions, there is even a one-to-one correspondence between Markovian invariant measures and stationary measures of the Markov semi-group [5, 31]. The inverse operation of $\mu \mapsto \rho = \mathbb{E}[\mu_\bullet]$ is then given by $\rho \mapsto \mu_\omega = \lim_{t\to\infty} \Phi(-t, \omega)^{-1}\rho$; the latter is in fact the pullback limit of ρ due to the cocycle property [31]. It follows readily from this result that RDS theory "sees" many more invariant measures than those given by the Markov semi-group approach: non-Markovian measures appear to play an important role in stochastic bifurcation theory [5], for instance.

To summarize, one might say that the classical forward approach considers only expectations and PDFs, whereas the RDS approach "slices" the statistics very finely: the former takes a hammer to the problem, while the latter takes a scalpel. Clearly, distinct physical processes may lead to the same observed PDF: the RDS approach and, in particular, the pullback limit are able to discriminate between these processes and thus provide further insight into them.

3.4 Time-Dependent Forcing and Ocean–Atmosphere Coupling

Given the NDS and RDS concepts of pullback attraction and time-dependent invariant measures, as introduced in the previous section, we are ready now to return to the wind-driven ocean circulation of Sect. 3.2 and consider two additional steps in its understanding, simulation and prediction, namely time-dependent forcing and coupling with the atmosphere above. We start by applying a still prescribed but now time-dependent wind-stress to a double-gyre model.

3.4.1 Time-Dependent Forcing

The most obvious way in which wind stress varies in mid-latitudes is from summer to winter: the so-called atmospheric jet stream is both stronger and closer to the equator in winter. Such periodic changes in the forcing were already considered in [155], albeit without the theoretical underpinnings discussed herein. The first application of pullback attractors to the double-gyre problem appears in [124], for the case of periodic forcing, and we rely here on [125] for a systematic presentation of this novel application of NDS theory to the wind-driven ocean circulation.

Model Formulation

The model, in its autonomous form [123], differs somewhat from that of Eq. (101) by the absence of the Laplacian eddy viscosity—given by the bilaplacian $\Delta^2 \psi$ of the streamfunction ψ—and by a slightly different form of the wind stress. Its discretized form is obtained by projection of Eqs. (101) onto a set of cartesian basis functions, and low-order truncation of the expansion thus obtained. The discrete, low-order model is thus governed by four nonlinear coupled ODEs for the variables $\{\Psi_i(t), i = 1, \cdots, 4\}$ that are the coefficients of the streamfunction $\psi(x, y, t)$ retained in this expansion. The choice of basis functions follows up on such idealized double-gyre models introduced in [76, 143], in particular on this basis including the exponential form of a current that decays away from the domain's western boundary.

In the presence of nonautonomous forcing, the model takes the form

$$
\begin{aligned}
\dot{\Psi}_1 + L_{11}\Psi_1 + L_{13}\Psi_3 + B_1(\Psi, \Psi) &= W_1(t), \\
\dot{\Psi}_2 + L_{22}\Psi_2 + L_{24}\Psi_4 + B_2(\Psi, \Psi) &= W_2(t), \\
\dot{\Psi}_3 + L_{33}\Psi_3 + L_{31}\Psi_1 + B_3(\Psi, \Psi) &= W_3(t), \\
\dot{\Psi}_4 + L_{44}\Psi_4 + L_{42}\Psi_2 + B_4(\Psi, \Psi) &= W_4(t).
\end{aligned}
\tag{109}
$$

Here Ψ denotes the vector of expansion coefficients $(\Psi_1, \Psi_2, \Psi_3, \Psi_4)$ of the streamfunction, $\mathbf{W}$ is the vector of forcing terms (W_1, W_2, W_3, W_4), and the bilinear terms B_i are given by

$$B_1(\Psi, \Psi) = 2J_{112}\Psi_1\Psi_2 + 2J_{114}\Psi_1\Psi_4 + 2J_{123}\Psi_2\Psi_3 + 2J_{134}\Psi_3\Psi_4,$$

$$B_2(\Psi, \Psi) = J_{211}\Psi_1^2 + J_{222}\Psi_2^2 + J_{233}\Psi_3^2 + J_{244}\Psi_4^2$$
$$+ 2J_{213}\Psi_1\Psi_3 + 2J_{224}\Psi_2\Psi_4,$$

$$B_3(\Psi, \Psi) = 2J_{312}\Psi_1\Psi_2 + 2J_{314}\Psi_1\Psi_4 + 2J_{323}\Psi_2\Psi_3 + 2J_{334}\Psi_3\Psi_4,$$

$$B_4(\Psi, \Psi) = J_{411}\Psi_1^2 + J_{422}\Psi_2^2 + J_{433}\Psi_3^2 + J_{444}\Psi_4^2 \tag{110}$$
$$+ 2J_{413}\Psi_1\Psi_3 + 2J_{424}\Psi_2\Psi_4.$$

In [125], the forcing $\mathbf{W}(\mathbf{x}, t)$ is defined as

$$\mathbf{W}(\mathbf{x}, t) = \gamma\big[1 + \epsilon f(t)\big]\mathbf{w}(\mathbf{x}), \tag{111}$$

where $\mathbf{x} = (x, y)$ and $\mathbf{w}(\mathbf{x})$ is the time-constant double-gyre wind stress curl used in [123]; furthermore, γ is the dimensionless intensity of this wind stress, while $f(t)$ is an aperiodic time dependence weighed by the dimensionless parameter ϵ. Clearly, $\mathbf{w}(\mathbf{x})$ is projected onto the same four leading-order basis functions as Ψ and $\mathbf{W}$.

Model Behavior

The aperiodic forcing $f(t)$ in Fig. 20a corresponds to an idealized but aperiodic choice with a substantial decadal component. Such decadal climate signals are associated with the so-called Pacific Decadal Oscillation [21, 104], for instance, and climate prediction on such time scales is a matter of great recent interest for the climate sciences [20, 61, 73, 74, 126].

The evolution of the model solutions subject to this forcing is shown in Fig. 20b, c for two reference cases, $\gamma = 0.96$ and $\gamma = 1.1$, respectively; the solutions of the corresponding autonomous system, i.e. with $\epsilon = 0$, are plotted in [125, Figs. 1(b,c)]. It is clear from the two panels that a regime change occurs as the forcing is increased. This regime change occurs in the autonomous system at $\gamma = 1.0$ and corresponds to a global bifurcation associated with a homoclinic orbit; see [123, 143] and, here, Paragraph on Global bifurcations. The persistence of the regime shift in the nonautonomous case, for ϵ as large as 0.2, is interesting but none too surprising.

In Fig. 20b, c, N trajectories $\Psi_k(t)$ emanate from N initial states uniformly distributed at time $t_0 = 0$ on a reference subset Γ of the (Ψ_1, Ψ_3)-plane, while the (Ψ_2, Ψ_4) coordinates of the initial states (not shown) are chosen the same way as in [124]. The panels (b, c) here provide a first representation of the

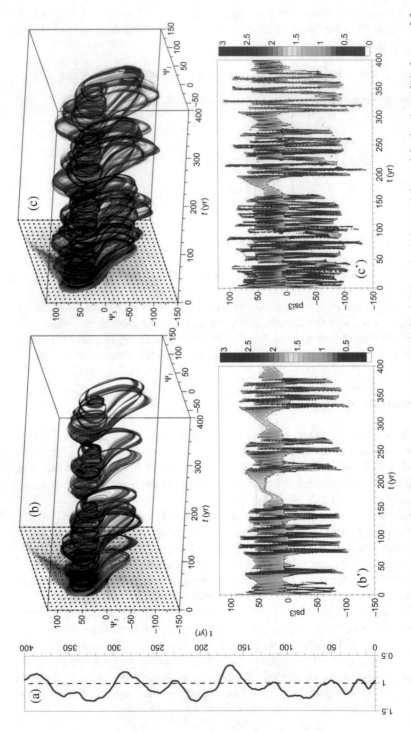

Fig. 20 Ensemble behavior of forced solutions of the double-gyre ocean model of [125]. (**a**) Time dependence of the total forcing $1 + \epsilon f(t)$, for $\epsilon = 0.2$. (**b**, **c**) Evolution of $N = 644$ initial states emanating from the subset Γ in the (Ψ_1, Ψ_3)-plane for (**b**) $\gamma = 0.96$ and (**c**) $\gamma = 1.1$. (**b'**, **c'**) Corresponding time series of $P\psi_3$. Reproduced from [125], with the permission of the American Meteorological Society

sets that approximate the corresponding PBAs; the small number of trajectories is in fact chosen for the sake of graphical clarity. The correct identification and characterization of the PBAs requires, however, an analysis of the PDFs that evolve along the trajectories and of the convergence to the appropriate invariant time-dependent sets; see Fig. 20b', c'. This convergence was shown in [125] to take no longer than about 15 years of model simulation.

Multiplicity of PBAs

The existence of a global PBA is rigorously demonstrated for the weakly dissipative, nonlinear model governed by Eqs. (109)–(111) in [125, Appendix], based on general, NDS-theoretical results [16, 83]. Numerically, though, this unique global attractor seems to possess two separate local PBAs, as apparent from Fig. 21. Panels (a) and (b) in the figure refer again to the two types of regimes apparent in Fig. 20b, c, for $\gamma = 0.96$ and $\gamma = 1.1$, respectively.

The mean normalized distance plotted in the figure is defined as follows:

$$\sigma(\Psi_1, \Psi_3) = \langle \delta_n \rangle T_{\text{tot}}. \tag{112}$$

Here δ_t is the distance, at time t, between two trajectories of the model that were a distance δ_0 apart at time $t = t_0$, and the normalized distance $\delta_n = \delta_t / \delta_0$ is averaged over the whole forward time integration $T_{\text{tot}} = t_{\text{fin}} - t_{\text{init}}$ of the available trajectories.

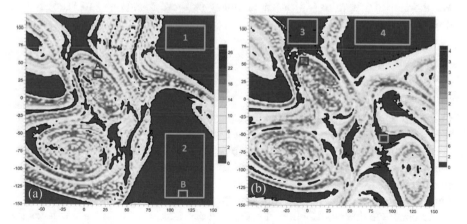

Fig. 21 Mean normalized distance $\sigma(\Psi_1, \Psi_3)$ for 15,000 trajectories of the double-gyre ocean model starting in the initial set Γ: (**a**) $\gamma = 0.96$, and (**b**) $\gamma = 1.1$. Reproduced from [125], with the permission of the American Meteorological Society

The maps of σ in Fig. 21 reveal large chaotic regions where $\delta_n \gg 1$ on average (warm colors) but also non-chaotic regions, in which $\sigma \leq 1$ (blue) and thus initially close trajectories do remain close on average. The rectangular regions in the two panels that are labeled by letters A and B and by numbers 1–4 correspond to subdomains of the initial set Γ and are discussed further in [125, Sec. 5]. The numerical evidence in Fig. 21 suggests, furthermore that the boundary between the two types of local attractors has fractal properties.

In the autonomous context, the coexistence of topologically distinct local attractors is well known in the climate sciences [37, 39, 56, 143, 144, and references therein]. The coexistence of local PBAs with chaotic vs. non-chaotic characteristics, within a unique global PBA, as illustrated by Fig. 21 here, seems to be novel, at least in the climate literature.

3.4.2   Coupled Ocean–Atmosphere Variability

Forcing of an ocean model by time-dependent wind stress is but one step in the direction of a fully coupled ocean–atmosphere model. Simple models of this type have been considered, for instance, by S. Vannitsem and associates [165, 167, 168], who also included heat fluxes as an important ingredient in the coupling.

A key conclusion of this group's work is that, in such a simple coupled model, it is possible to find a fairly broad parameter regime within which slow modes, with interdecadal near-periodicities exist. Moreover, these modes are truly coupled, in the sense that their atmospheric components share the long near-periodicities. We merely refer here to [55] for a summary of these results and to [166] for an observational verification.

3.5   *Concluding Remarks*

Over the last four decades, the climate sciences have become of paramount importance in understanding, predicting and attempting to improve the relations between humanity and the planet it inhabits [71–74, 114]. This development could not have occurred without a strong contribution from the mathematical sciences. Several books [37, 38, 56, 97, 102, 105, 164] have documented the mathematics that have been used and that have in turn benefited from this area of applications.

To conclude this contribution, it seems appropriate to formulate a few ideas for possibly fruitful future applications of NDS and RDS theory to the climate sciences. It would appear to the authors that—in spite of increasing efforts in this direction, e.g. [107, and further articles in the same volume]—insufficient attention has been paid to understanding the strong interaction between natural or intrinsic climate variability [20, 115], on the one hand, and the anthropogenic forcing upon which a huge fraction of the recent climate literature has focused [73, 74]. We would like, therefore, to suggest an avenue for overcoming this gap, and to summon the necessary mathematics for doing so.

3.5.1 A Generalized Definition of Climate Sensitivity

The usual way of defining *climate sensitivity*, going back to [114], relies on the conceptual picture of Fig. 22a: climate is in a steady state characterized by a global average temperature $\overline{T} = \overline{T}_0$; a parameter that is affected by human activities, say CO_2 concentration, is suddenly changed; and $\overline{T}(t)$ follows—according to the solution of a scalar, linear ODE like Eq. (105), in which $\overline{T}$ replaces x and a Heaviside function replaces σt on the right-hand side—so that $\overline{T}(t) \to \overline{T}_1$, and $\overline{T}_1 > \overline{T}_0$ when the CO_2 jump is positive.

This cartoon has been, of course, considerably enriched in successive assessment reports of the the Intergovernmental Panel on Climate Change [71–74] from a scalar, linear ODE to the coupled, nonlinear PDE systems that govern GCMs. But no mathematically consistent picture has emerged for reconciling the intrinsic variability that those PDE systems include, as illustrated in Sects. 3.2 and 3.4 herein, with the complex nature of the forcing, both natural and anthropogenic. Thus climate sensitivity is still essentially thought of as $\Delta\overline{T}/\Delta CO_2$, where $\Delta\overline{T} = \overline{T}_1 - \overline{T}_0$, the difference between a final and an initial "equilibrium" temperature, and ΔCO_2 is likewise the difference between a final and an initial forcing level. Consideration has been given to changes in other fields than temperature, such as precipitation [95, 133], to variances and energetics [98, 99], or to regional differences [14, 100].

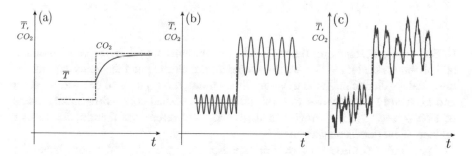

Fig. 22 Climate sensitivity for (**a**) an equilibrium model; and (**b, c**) a nonequilibrium model. Given a jump in a parameter, such as CO_2 concentration, only the mean global temperature $\overline{T}$ changes in (**a**), while in (**b**) it is also the period, amplitude and phase of a purely periodic oscillation, such as the seasonal cycle or the intrinsic ENSO cycle. Finally, in panel (**c**), it is also the character of the oscillation, whether deterministic or stochastically perturbed, which may change. After [54]

To include these considerations and follow up on ideas formulated in [54], we would like to suggest a definition of climate sensitivity γ^3 that takes full advantage of the NDS and RDS framework and that encompasses changes in the natural variability, in its most general meaning, as the forcing level or some other parameter changes. This definition replaces the more-or-less standard definition of $\gamma = \partial \overline{T}/\partial \mu$ [56, p. 320], with μ an arbitrary parameter generalizing CO_2—a definition that corresponds to the equilibrium situation in Fig. 22a—by a more appropriate and self-consistent one that corresponds to the chaotic and random situation in Fig. 22c.

3.5.2 PBAs, Invariant Measures and Wasserstein Distance

The mathematical framework of PBAs, in the presence of time-dependent forcing or coefficients, suggests the definition

$$\gamma = \partial d_W/\partial \mu, \tag{113}$$

where d_W is the *Wasserstein distance* [41, 54] between the time-dependent invariant measures supported on the system's PBA at two distinct parameter values. An idea of the large differences that may exist between two such PBAs is given by the three snapshots of the invariant measure on the PBA of the infinite-dimensional, but still relatively simple ENSO model of [51] in Fig. 23 below.

The model's two dependent variables are sea surface temperature T in the eastern Tropical Pacific and thermocline depth h there, as a function of time t:

$$\dot{T} = f(T(t), h(t)), \quad h(t) = g(T, h, F)(t, \tau_1, \tau_2), \quad F(t) = 1 + \epsilon cos(\omega t + \phi). \tag{114}$$

In Eqs. (114), F stands for the seasonal forcing, with period $2\pi/\omega = 12$ months, and all three variables—T, h and F—depend on the time t and the delays τ_1 and τ_2; these delays characterize the traveling times along the Equator of eastward Kelvin and westward Rossby waves. Several authors have studied delay-differential models of ENSO; see [37] for a review and [27, 62, and references therein] for further mathematical details on such models.

The solutions of Eqs. (114) exhibit periodic, quasi-periodic and chaotic behavior, as well as frequency locking to the time-dependent, seasonal cycle. Thus, in principle, an infinite number of scalars are required to define the dependence of these solutions on the parameters τ_1 and τ_2; these scalars need to include not just the means of temperature T and depth h, but also their variance and higher-order moments. In [54, Fig. 7] this dependence was shown for the zeroth, second

---

[3]This γ should not be confused with the parameter used in Sect. 3.4.1 for the forcing intensity.

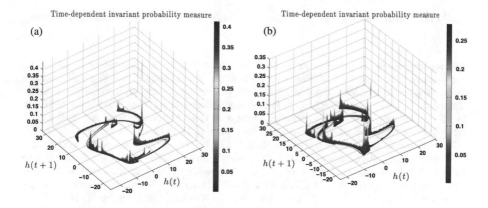

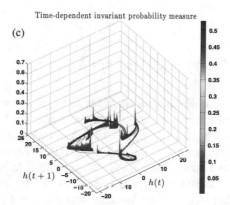

Fig. 23 Time-dependent invariant measures of the [51] model; snapshots shown at three times t: (**a**) $t_1 = 19.23$ year, (**b**) $t_2 = 20$ year and (**c**) $t_3 = 20.833$ year. Courtesy of M. D. Chekroun

and fourth moments of $h(t)$; more precisely the plotted quantities were the mean, standard deviation and fourth root of the kurtosis of the PDF.

While the figure illustrates successive snapshots of the invariant measure supported on the PBA for the same value of the model parameters, there are similarly large differences between such measures for different parameter values [54, 62]. Closer attention to the definition of Eq. (113) can provide better insights into the changes in time, as well as with respect to a parameter, of higher moments of a model's PDF but also into the delicate and important matter of changes in the distribution of extreme events [22, 66].

Steps in the application of the NDS and RDS framework to climate change and climate sensitivity are being taken by a number of research groups [10, 11, 38, 43, 100]; see especially [58] and references therein. This area of research is only opening up and should provide many opportunities for contributions by mathematicians and climate scientists, as well as for fruitful interactions among them.

Acknowledgements It is a pleasure to thank M. D. Chekroun, D. Kondrashov, H. Liu, J. C. McWilliams, J. D. Neelin and I. Zaliapin for many useful discussions and their continuing interest in the questions studied here. The research reported in Sect. 3.4 was carried out jointly with M. D. Chekroun, L. De Cruz, J. Demayer, S. Pierini and S. Vannitsem [125, 168]. Figure 23 is due to M. D. Chekroun, a steadfast companion on the road to understanding NDS and RDS theory, along with their applications to climate dynamics in general. H. Liu helped with finalizing Fig. 22. It is a pleasure to thank the organizers of the INdAM Workshop on *Mathematical Approaches to Climate Change Impacts*—P. Cannarsa, D. Mansutti, and A. Provenzale—for the invitation to deliver a lecture and to contribute to this volume. The writing of this research-and-review paper was supported by grants N00014-12-1-0911 and N00014-16-1-2073 from the Multidisciplinary University Research Initiative (MURI) of the Office of Naval Research and by the US National Science Foundation grant OCE 1243175. This paper is TiPES contribution #4; this project has received funding from the European Union Horizon 2020 research and innovation programme under grant agreement No. 820970.

Appendix 1: RDS Theory and Random Attractors

We present here briefly the mathematical concepts and tools of random dynamical systems, random attractors and stochastic equivalence. We shall use the concept of pullback attractor introduced in Sect. 3.2.1 to define the closely related notion of a random attractor, but need first to define an RDS. We denote by $\mathbb{T}$ the set $\mathbb{Z}$, for maps, or $\mathbb{R}$, for flows. Let $(X, \mathcal{B})$ be a measurable phase space, and $(\Omega, \mathcal{F}, \mathbb{P}, (\theta(t))_{t \in \mathbb{T}})$ be a *metric* dynamical system *i.e.* a flow in the probability space $(\Omega, \mathcal{F}, \mathbb{P})$, such that $(t, \omega) \mapsto \theta(t)\omega$ is measurable and $\theta(t) : \Omega \to \Omega$ is measure preserving, *i.e.*, $\theta(t)\mathbb{P} = \mathbb{P}$.

Let $\varphi : \mathbb{T} \times \Omega \times X \to X$, $(t, \omega, x) \mapsto \varphi(t, \omega)x$, be a mapping with the two following properties:

(R$_1$): $\varphi(0, \omega) = \mathrm{Id}_X$, and
(R$_2$) (the cocycle property): For all $s, t \in \mathbb{T}$ and all $\omega \in \Omega$,

$$\varphi(t + s, \omega) = \varphi(t, \theta(s)\omega) \circ \varphi(s, \omega).$$

If φ is measurable, it is called a *measurable* RDS over θ. If, in addition, X is a topological space (respectively a Banach space), and φ satisfies $(t, \omega) \mapsto \varphi(t, \omega)x$ continuous (resp. C^k, $1 \leq k \leq \infty$) for all $(t, \omega) \in \mathbb{T} \times \Omega$, then φ is called a *continuous* (resp. C^k) RDS over the flow θ. If so, then

$$(\omega, x) \mapsto \Theta(t)(x, \omega) := (\theta(t)\omega, \varphi(t, \omega)x), \qquad (115)$$

is a (measurable) flow on $\Omega \times X$, and is called the *skew-product* of θ and φ. In the sequel, we shall use the terms "RDS" or "cocycle" synonymously.

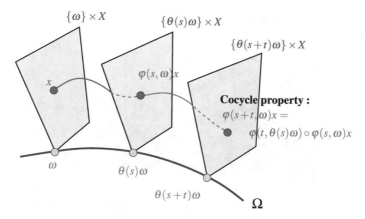

Fig. 24 Random dynamical systems (RDS) viewed as a flow on the bundle $X \times \Omega$ = "dynamical space" × "probability space." For a given state x and realization ω, the RDS φ is such that $\Theta(t)(x, \omega) = (\theta(t)\omega, \varphi(t, \omega)x)$ is a flow on the bundle. Redrawn after from [64], with permission from Elsevier

The choice of the so-called *driving system* θ is a crucial step in this set-up; it is mostly dictated by the fact that the coupling between the stationary driving and the deterministic dynamics should respect the time invariance of the former, as illustrated in Fig. 24. The driving system θ also plays an important role in establishing *stochastic conjugacy* [29] and hence the kind of classification we aim at.

The concept of *random attractor* is a natural and straightforward extension of the definition of pullback attractor (104), in which Sell's [137] process is replaced by a cocycle, cf. Fig. 24, and the attractor $\mathcal{A}$ now depends on the realization ω of the noise, so that we have a family of random attractors $\mathcal{A}(\omega)$, cf. Fig. 25. Roughly speaking, for a fixed realization of the noise, one "rewinds" the noise back to $t \rightarrow -\infty$ and lets the experiment evolve (forward in time) towards a possibly attracting set $\mathcal{A}(\omega)$; the driving system θ enables one to do this rewinding without changing the statistics, cf. Figs. 24 and 25.

Other notions of attractor can be defined in the stochastic context, in particular based on the original SDE; see [32, 33] for a discussion on this topic. The present definition, though, will serve us well.

Having defined RDSs and random attractors, we now introduce the notion of *stochastic equivalence* or *conjugacy*, in order to rigorously compare two RDSs; it is defined as follows: two cocycles $\varphi_1(\omega, t)$ and $\varphi_2(\omega, t)$ are conjugated if and only if there exists a random homeomorphism $h \in \text{Homeo}(X)$ and an invariant set such that $h(\omega)(0) = 0$ and

$$\varphi_1(\omega, t) = h(\theta(t)\omega)^{-1} \circ \varphi_2(\omega, t) \circ h(\omega). \tag{116}$$

Stochastic equivalence extends classic topological conjugacy to the bundle space $X \times \Omega$, stating that there exists a one-to-one, stochastic change of variables that

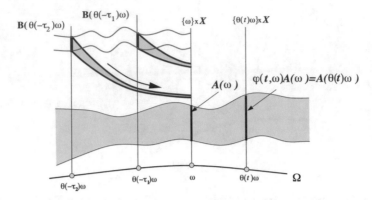

Pullback attraction to A(ω)

Fig. 25 Schematic diagram of a random attractor $\mathcal{A}(\omega)$, where $\omega \in \Omega$ is a fixed realization of the noise. To be attracting, for every set B of X in a family $\mathfrak{B}$ of such sets, one must have $\lim_{t \to +\infty} \text{dist}(B(\theta(-t)\omega), \mathcal{A}(\omega)) = 0$ with $B(\theta(-t)\omega) := \varphi(t, \theta(-t)\omega)B$; to be invariant, one must have $\varphi(t, \omega)\mathcal{A}(\omega) = \mathcal{A}(\theta(t)\omega)$. This definition depends strongly on $\mathfrak{B}$; see [32] for more details. Reproduced from [64], with permission from Elsevier

continuously transforms the phase portrait of one sample system in X into that of any other such system.

Appendix 2: Mixing in Random Dynamical Systems

In this appendix, we define rigorously the concept of an ω-wise mixing RDS, in the continuous-time context. Recall first the well-known definition of mixing in a deterministic dynamical system. Given a flow $\{\phi_t\}$ on a topological space X, which possesses an invariant (Borel) probability measure μ, we say that the dynamical system (ϕ_t, μ) is mixing if for any two measurable sets A and B,

$$\mu(A \cap \phi_{-t}(B)) \underset{t \to \infty}{\longrightarrow} \mu(A)\mu(B), \tag{117}$$

or equivalently,

$$\int F \cdot (G \circ \phi_t) d\mu \underset{t \to \infty}{\longrightarrow} \int F d\mu \int G d\mu, \tag{118}$$

for any pair of continuous functions $F, G : X \to \mathbb{R}$. Equation (117) states that the set of points in A whose images belong to B by $\{\phi_t\}$ tends towards having the same proportion in A as B has in X, with proportions being understood in terms of the measure μ. Hence any measurable set will tend to redistribute itself over the state space according to μ.

Let us now consider a cocycle $\{\Phi(t, \omega)\}_{(t,\omega)\in\mathbb{R}\times\Omega}$ on the base space $(\Omega, \mathcal{F}, \mathbb{P}, \{\theta_t\})$, which possesses the sample measures $\{\mu_\omega\}$. We say that Φ is ω-wise mixing or fiber mixing [12]—or even simply mixing, if no confusion is possible—if for any random sets [32] $A(\omega)$ and $B(\omega)$,

$$\mu_\omega\Big(A(\omega) \cap \Phi(t, \omega)^{-1}\big(B(\theta_t\omega)\big)\Big) \xrightarrow[t\to\infty]{} \mu_\omega(A(\omega))\mu_{\theta_t\omega}(B(\theta_t\omega)), \qquad (119)$$

almost surely with respect to $\mathbb{P}$. This mixing concept and its interpretation are natural extensions of their deterministic counterparts just recalled above, except that the mixing property has to be checked across the fibers ω and $\theta_t\omega$, due to the skew-product nature of the RDS (Φ, θ) [12].

Appendix 3: Low-Frequency Variability (LFV) and Mixing

Low-frequency variability (LFV) is a widely used, but not clearly defined concept in the atmospheric, oceanic and climate sciences [56, 59, 64]. In general, one just refers to phenomena whose periods are longer than those previously studied. Examples include atmospheric LFV—referring to so-called intraseasonal oscillations whose characteristic time scale of 10–100 days is longer than the 5–10-day life cycle of mid-latitude storms but not longer than a season [59, 60]—or oceanic LFV, referring to interannual or interdecadal variability whose characteristic time scales are longer than the several-months–long ones of mesoscale eddies and the seasonal cycle of a year [53, 59].

In this appendix, we clarify the notion of LFV from a mathematical perspective. Let us reconsider the deterministic Lorenz system [96]. It is known that the power spectral density, or power spectrum, of this system is exponentially decaying [57, 116]. At the same time, one can check numerically that the decay of the auto-correlation function is exponentially decaying, too. Other types of power-spectrum behavior may be encountered for chaotic dynamical systems, though. Aside from pure power-law decay, it may also happen that the power spectrum contains one or several broad peaks that stand out above the continuous background, whether the latter has a power-law [9] or exponential decay. If the central frequencies of these peaks are located in a frequency band that lies close to the lower end of the frequency range being studied, the system is said to exhibit LFV [57, 63].

This climatically motivated, but vague notion of LFV can be formalized mathematically through the mixing concept introduced in Appendix 2. Indeed, for a general flow $\{\phi_t\}$ on a topological space X, which possesses an invariant (Borel) probability measure μ, let us define the correlation function by

$$C_t(F, G) := \left| \int F \cdot (G \circ \phi_t) d\mu - \int F d\mu \int G d\mu \right|,$$

using the same notations as above. If the system (ϕ_t, μ) is mixing, the rate of approach to zero of $C_t(F, G)$ is called the rate of decay of correlations for its observables F and G. A system exhibits a *slow decay rate of correlations at short lags* if the rate is slower than exponential over some characteristic time interval $[0, T]$.

This emphasis on the nonuniform decay rate of correlations is consistent with the heuristic LFV notion used in the climate sciences, as described above, and it connects the mixing properties of the flow with its power spectral density. In [25], the authors elucidated the relationship between these two approaches, as well as the relationships between the rate of decay of correlations and the occurrence of rough or smooth parameter dependence of the model's statistics. Moreover, the quantitative knowledge about the system's LFV—when combined with the pathwise approach discussed in Sect. 3.3.2—can be used, according to [24, 86], to improve ensemble prediction skills of data-driven stochastic models, such as those discussed in [87].

References

1. Allen, M.R.: Do-it-yourself climate prediction. Nature **401**, 627 (1999)
2. Andronov, A.A., Pontryagin, L.S.: Systèmes grossiers. Dokl. Akad. Nauk. SSSR **14**(5), 247–250 (1937)
3. Araujo, V., Pacifico, M., Pujal, R., Viana, M.: Singular-hyperbolic attractors are chaotic. Trans. Amer. Math. Soc. **361**, 2431–2485 (2009)
4. Arnold, V.I.: Geometrical Methods in the Theory of Differential Equations, 334 pp. Springer, Berlin (1983)
5. Arnold, L.: Random Dynamical Systems, 616 pp. Springer, Berlin (1998)
6. Bell, D.R.: Degenerate Stochastic Differential Equations and Hypoellipticity. Longman, Harlow (1995)
7. Berger, A., Siegmund, S.: On the gap between random dynamical systems and continuous skew products. J. Dyn. Diff. Equ. **15**, 237–279 (2003)
8. Berloff, P., Hogg, A., Dewar, W.: The turbulent oscillator: a mechanism of low-frequency variability of the wind-driven ocean gyres. J. Phys. Oceanogr. **37**, 2363–2386 (2007)
9. Bhattacharya, K., Ghil, M., Vulis, I.L.: Internal variability of an energy-balance model with delayed albedo effects. J. Atmos. Sci. **39**, 1747–1773 (1982). https://doi.org/10.1175/1520-0469
10. Bódai, T., Károlyi, G., Tél, T.: A chaotically driven model climate: extreme events and snapshot attractors. Nonlin. Processes Geophys. **18**, 573–580 (2011)
11. Bódai, T., Lucarini, V., Lunkeit, F., Boschi, R.: Global instability in the Ghil-Sellers model. Clim. Dyn. **44**, 3361–3381 (2015)
12. Bogenschütz, T., Kowalski, Z.S.: A condition for mixing of skew products. Aequationes Math. **59**, 222–234 (2000)
13. Bouchet, F., Sommeria, J.: Emergence of intense jets and Jupiter's great red spot as maximum entropy structures. J. Fluid. Mech. **464**, 165–207 (2002)
14. Bracco, A., Neelin, J.D., Luo, H., McWilliams, J.C., Meyerson, J.E.: High-dimensional decision dilemmas in climate models. Geosci. Model Dev. **6**, 1673–1687 (2013). https://doi.org/10.5194/gmdd-6-2731-2013
15. Carvalho, A.N., Langa, J.A., Robinson, J.C.: Lower semicontinuity of attractors for non-autonomous dynamical systems. Ergod. Theory Dyn. Syst. **29**, 765–780 (2009)

16. Carvalho, A., Langa, J.A., Robinson, J.C.: Attractors for Infinite-Dimensional Non-Autonomous Dynamical Systems, 391 pp. Springer, Berlin (2012)

17. Cessi, P., Ierley, G.R.: Symmetry-breaking multiple equilibria in quasigeostrophic wind-driven flows. J. Phys. Oceanogr. **25**, 1196–1205 (1995)

18. Chang, P., Ji, L., Li, H., Flugel, M.: Chaotic dynamics versus stochastic processes in El Niño-Southern Oscillation in coupled ocean-atmosphere models. Physica D **98**, 301–320 (1996)

19. Chang, K.I., Ide, K., Ghil, M., Lai, C.C.A.: Transition to aperiodic variability in a wind-driven double-gyre circulation model. J. Phys. Oceanogr. **31**, 1260–1286 (2001)

20. Chang, C.P., Ghil, M., Latif, M., Wallace, J.M.: Climate Change: Multidecadal and Beyond, vol. 6, 388 pp. World Scientific Publishing Co./Imperial College Press (2015)

21. Chao, Y., Ghil, M., McWilliams, J.C.: Pacific interdecadal variability in this century's sea surface temperatures. Geophys. Res. Lett. **27**, 2261–2264 (2000)

22. Chavez, M., Ghil, M., Urrutia Fucugauchi, J.: Extreme Events: Observations, Modeling and Economics, vol. 214, 438 pp. American Geophysical Union/Wiley, Washington/Hoboken (2015)

23. Chekroun, M.D., Simonnet, E., Ghil, M.: Stochastic climate dynamics: random attractors and time-dependent invariant measures. Physica D **240**, 1685–1700 (2011)

24. Chekroun, M.D., Kondrashov, D., Ghil, M.: Predicting stochastic systems by noise sampling, and application to the El Niño-Southern Oscillation. Proc. Natl. Acad. Sci. USA **108**, 11766–11771 (2011)

25. Chekroun, M.D., Neelin, J.D., Kondrashov, D., McWilliams, J.C., Ghil, M.: Rough parameter dependence in climate models: the role of Ruelle-Pollicott resonances. Proc. Natl. Acad. Sci. USA **111**, 1684–1690 (2014). https://doi.org/10.1073/pnas.1321816111

26. Chekroun, M.D., Liu, H., Wang, S.: Approximation of Stochastic Invariant Manifolds: Stochastic Manifolds for Nonlinear SPDEs I. Springer Briefs in Mathematics. Springer, Berlin (2015)

27. Chekroun, M.D., Ghil, M., Liu, H., Wang, S.: Low-dimensional Galerkin approximations of nonlinear delay differential equations. Discr. Cont. Dyn. Syst. **36**(8), 4133–4177 (2016)

28. Collet, P., Tresser, C.: Ergodic theory and continuity of the Bowen-Ruelle measure for geometrical flows. Fyzika **20**, 33–48 (1988)

29. Cong, N.D.: Topological Dynamics of Random Dynamical Systems. Oxford Mathematical Monographs, 212 pp. Clarendon Press, Oxford (1997)

30. Crauel, H.: White noise eliminates instability. Arch. Math. **75**, 472–480 (2000)

31. Crauel, H.: Random probability measures on Polish spaces, vol. 11. Stochastic Monographs. Taylor & Francis, Didcot (2002)

32. Crauel, H.: A uniformly exponential attractor which is not a pullback attractor. Arch. Math. **78**, 329–336 (2002)

33. Crauel, H., Flandoli, F.: Attractors for random dynamical systems. Technical Report 148, cuola Normale Superiore Pisa (1992)

34. Crauel, H., Flandoli, F.: Additive noise destroys a pitchfork bifurcation. J. Dyn. Diff. Equ. **10**, 259–274 (1998)

35. Cushman-Roisin, B., Beckers, J.-M.: Introduction to Geophysical Fluid Dynamics: Physical and Numerical Aspects, 2nd edn., 875 pp. Academic, Cambridge (2011)

36. Da Costa, E.D., Colin de Verdière, A.C.: The 7.7 year North Atlantic oscillation. Q. J. R. Meteorol. Soc. **128**, 797–818 (2004)

37. Dijkstra, H.A.: Nonlinear Physical Oceanography: A Dynamical Systems Approach to the Large Scale Ocean Circulation and El Niño, 2nd edn., 532 pp. Springer, Berlin (2005)

38. Dijkstra, H.A.: Nonlinear Climate Dynamics, 367 pp. Cambridge University Press, Cambridge (2013)

39. Dijkstra, H.A., Ghil, M.: Low-frequency variability of the large-scale ocean circulation: a dynamical systems approach. Rev. Geophys. **43** (2005). https://doi.org/10.1029/2002RG000122

40. Dijkstra, H.A., Katsman, C.A.: Temporal variability of the wind-driven quasi-geostrophic double gyre ocean circulation: basic bifurcation diagrams. Geophys. Astrophys. Fluid Dyn. **85**, 195–232 (1997)

41. Dobrushin, R.L.: Prescribing a system of random variables by conditional distributions. Theor. Prob. Appl. **15**, 458–486 (1979)
42. Dorfle, M., Graham, R.: Probability density of the Lorenz model. Phys. Rev. A **27**, 1096–1105 (1983)
43. Drótos, G., Bódai, T., Tél, T.: Probabilistic concepts in a changing climate: a snapshot attractor picture. J. Clim. **28**, 3275–3288 (2015)
44. Dubar, M.: Approche climatique de la période romaine dans l'est du Var: recherche et analyse des composantes périodiques sur un concrétionnement centennal (Ier-IIe siècle apr. J.-C.) de l'aqueduc de Fréjus. Archeoscience **30**, 163–171 (2006)
45. Eckmann, J.P., Ruelle, D.: Ergodic theory of chaos and strange attractors. Rev. Mod. Phys. **57**, 617–656 (1985)
46. Farrel, B.F., Ioannou, P.J.: Structural stability of turbulent jets. J. Atmos. Sci. **60**, 2101–2118 (2003)
47. Feliks, Y., Ghil, M., Simonnet, E.: Low-frequency variability in the mid-latitude atmosphere induced by an oceanic thermal front. J. Atmos. Sci. **61**, 961–981 (2004)
48. Feliks, Y., Ghil, M., Simonnet, E.: Low-frequency variability in the mid-latitude baroclinic atmosphere induced by an oceanic thermal front. J. Atmos. Sci. **64**, 97–116 (2007)
49. Feliks, Y., Ghil, M., Robertson, A.W.: Oscillatory climate modes in the eastern Mediterranean and their synchronization with the North Atlantic Oscillation. J. Clim. **23**, 4060–4079 (2010). https://doi.org/10.1175/2010JCLI3181.1
50. Feliks, Y., Ghil, M., Robertson, A.W.: The atmospheric circulation over the North Atlantic as induced by the SST field. J. Clim. **24**, 522–542 (2011). https://doi.org/10.1175/2010JCLI3859.1
51. Galanti, E., Tziperman, E.: ENSO's phase locking to the seasonal cycle in the fast-SST, fast-wave, and mixed-mode regimes. J. Atmos. Sci. **57**, 2936–2950 (2000)
52. Ghil, M.: Cryothermodynamics: the chaotic dynamics of paleoclimate. Physica D **77**, 130–159 (1994)
53. Ghil, M.: Hilbert problems for the geosciences in the 21st century. Nonlinear Process. Geophys. **8**, 211–222 (2001)
54. Ghil, M.: A mathematical theory of climate sensitivity or, How to deal with both anthropogenic forcing and natural variability?. In: Chang, C.P., Ghil, M., Latif, M., Wallace, J.M. (Eds.) Climate Change: Multidecadal and Beyond, pp. 31–51. World Scientific Publishing Co./Imperial College Press, Singapore/London (2015)
55. Ghil, M.: The wind-driven ocean circulation: applying dynamical systems theory to a climate problem. Discr. Cont. Dyn. Syst. – A **37**, 189–228 (2017). https://doi.org/10.3934/dcds.2017008
56. Ghil, M., Childress, S.: Topics in Geophysical Fluid Dynamics: Atmospheric Dynamics, Dynamo Theory, and Climate Dynamics, 512 pp. Springer, Berlin (1987)
57. Ghil, M., Jiang, N.: Recent forecast skill for the El Niño/Southern Oscillation. Geophys. Res. Lett. **25**, 171–174 (1998)
58. Ghil, M., Lucarini, V.: The physics of climate variability and climate change. Rev. Mod. Phys., 1–84 (2019). Submitted. arXiv:1910.00583
59. Ghil, M., Robertson, A.W.: Solving problems with GCMs: general circulation models and their role in the climate modeling hierarchy. In: Randall, D. (ed.) General Circulation Model Development: Past, Present and Future, pp. 285–325. Academic, San Diego (2000)
60. Ghil, M., Robertson, A.W.: 'Waves" vs "particles" in the atmosphere's phase space: a pathway to long-range forecasting? Proc. Natl. Acad. Sci. USA **99**, 2493–2500 (2002)
61. Ghil, M., Vautard, R.: Interdecadal oscillations and the warming trend in global temperature time series. Nature **350**, 324–327 (1991)
62. Ghil, M., Zaliapin, I.: Understanding ENSO variability and its extrema: a delay differential equation approach, vol. 214, ch. 6, pp. 63–78. In: Chavez, M., Ghil, M., Urrutia-Fucugauchi, J. (eds.) Extreme Events: Observations, Modeling and Economics. American Geophysical Union/Wiley, Washington/Hoboken (2015)

63. Ghil, M., Allen, M.R., Dettinger, M.D., Ide, K., Kondrashov, D., Mann, M.E., Robertson, A.W., Saunders, A., Tian, Y., Varadi, F., Yiou, P.: Advanced spectral methods for climatic time series. Rev. Geophys. **40**(1), 3.1–3.41 (2002)
64. Ghil, M., Chekroun, M.D., Simonnet, E.: Climate dynamics and fluid mechanics: natural variability and related uncertainties. Physica D **237**, 2111–2126 (2008)
65. Ghil, M., Zaliapin, I., Thompson, S.: A delay differential model of ENSO variability: parametric instability and the distribution of extremes. Nonlin. Processes Geophys. **15**, 417–433 (2008)
66. Ghil, M., Yiou, P., Hallegatte, S., Malamud, B.D., Naveau, P., Soloviev, A., Friederichs, P., Keilis-Borok, V., Kondrashov, D., Kossobokov, V., Mestre, O., Nicolis, C., Rust, H., Shebalin, P., Vrac, M., Witt, A., Zaliapin, I.: Extreme events: dynamics, statistics and prediction. Nonlin. Processes Geophys. **18**, 295–350 (2011). https://doi.org/10.5194/npg-18-295-2011
67. Gill, A.E.: Atmosphere-ocean dynamics, 662 pp. Academic, Cambridge (1982)
68. Guckenheimer, J., Williams, R.F.: Structural stability of Lorenz attractors. Publ. Math. I.H.E.S. **50**, 59–72 (1979)
69. Guckenheimer, J., Holmes, P.: Nonlinear Oscillations, Dynamical Systems and Bifurcations of Vector Fields, 2nd edn., 453 pp. Springer, Berlin (1991)
70. Holton, J., Hakim, G.J.: An Introduction to Dynamic Meteorology, 5th edn., 552 pp. Academic, Cambridge (2012)
71. IPCC: Climate change. In: Houghton, J.T., Jenkins, G.J., Ephraums, J.J. (eds.) The IPCC Scientific Assessment, 365 pp. Cambridge University Press, Cambridge (1991)
72. IPCC: Climate change 2001: the scientific basis. In: Houghton, J.T., Ding, Y., Griggs, D.J., Noguer, M., van der Linden, P.J., Dai, X., Maskell, K., Johnson, C.A. (eds.) Contribution of Working Group I to the Third Assessment Report of the Intergovernmental Panel on Climate Change (IPCC), 944 pp. Cambridge University Press, Cambridge (2001)
73. IPCC: Climate change 2007: the physical science basis. In: Solomon, S., Qin, D., Manning, M., Chen, Z., Marquis, M., Averyt, K.B., Tignor, M., Miller, H.L. (eds.) Contribution of Working Group I to the Fourth Assessment Report of the IPCC. Cambridge University Press, Cambridge (2007)
74. IPCC: Climate change 2013. In: Stocker, T.F., Qin, D., Plattner, G.K., Tignor, M., Allen, S.K., Boschung, J., Nauels, A., Xia, Y., Bex, B., Midgley, B.M. (eds.) The Physical Science Basis: Contribution of Working Group I to the Fifth Assessment Report of the Intergovernmental Panel on Climate Change. Cambridge University Press, Cambridge (2013)
75. Jiang, S., Jin, F.-F., Ghil, M.: The nonlinear behavior of western boundary currents in a wind-driven, double-gyre, shallow-water model, pp. 64–67. In: Ninth Conference Atmospheric & Oceanic Waves and Stability, San Antonio. American Meteorological Society, Boston (1993)
76. Jiang, S., Jin, F.-F., Ghil, M.: Multiple equilibria, periodic, and aperiodic solutions in a wind-driven, double-gyre, shallow-water model. J. Phys. Oceanogr. **25**, 764–786 (1995)
77. Jin, F.-F., Neelin, J.D., Ghil, M.: El Niño on the Devil's staircase: annual subharmonic steps to chaos. Science **264**, 70–72 (1994)
78. Jin, F.-F., Neelin, J.D., Ghil, M.: El Niño/Southern Oscillation and the annual cycle: subharmonic frequency locking and aperiodicity. Physica D **98**, 442–465 (1996)
79. Kalnay, E., Atmospheric Modeling, Data Assimilation and Predictability, 341 pp. Cambridge University Press, Cambridge (2003)
80. Katok, A., Haselblatt, B.: Introduction to the Modern Theory of Dynamical Systems, vol. 54, 822 pp. Encyclopedia of Mathematics and Its Applications. Cambridge University Press, Cambridge (1995)
81. Katsman, C.A., Dijkstra, H.A., Drijfhout, S.S.: The rectification of the wind-driven ocean circulation due to its instabilities. J. Mar. Res. **56**, 559–587 (1998)
82. Kifer, Y.: Ergodic Theory of Random Perturbations. Birkhäuser, Basel (1988)
83. Kloeden, P.E., Rasmussen, M.: Nonautonomous Dynamical Systems, vol. 176. Mathematical Surveys and Monographs. American Mathematical Society, Providence (2011)
84. Kohn, J.J.: Pseudo-differential operators and hypoellipticity. Proc. Amer. Math. Soc. Symp. Pure Math. **23**, 61–69 (1973)

85. Kondrashov, D., Feliks, Y., Ghil, M.: Oscillatory modes of extended Nile River records (A.D. 622–1922). Geophys. Res. Lett. **32**, L10702 (2005). https://doi.org/10.1029/2004GL022156
86. Kondrashov, K., Chekroun, M.D., Robertson, A.W., Ghil, M.: Low-order stochastic model and "past-noise forecasting" of the Madden-Julian oscillation. Geophys. Res. Lett. **40**, 5303–5310 (2013)
87. Kondrashov, D., Chekroun, M.D., Ghil, M.: Data-driven non-Markovian closure models. Physica D **297**, 33–55 (2015)
88. Kravtsov, S., Berloff, P., Dewar, W.K., Ghil, M., McWilliams, J.C.: Dynamical origin of low-frequency variability in a highly nonlinear mid-latitude coupled model. J. Climate **19**, 6391–6408 (2007)
89. Langa, J.A., Robinson, J.C., Suarez, A.: Stability, instability, and bifurcation phenomena in non-autonomous differential equations. Nonlinearity **15**, 887–903 (2002)
90. Lasota, A., Mackey, M.C.: Chaos, Fractals, and Noise: Stochastic Aspects of Dynamics, vol. 97. Applied Mathematical Sciences. Springer, Berlin (1994)
91. Ledrappier, F., Young, L.-S.: Entropy formula for random transformations. Probab. Theory Relat. Fields **80**, 217–240 (1988)
92. Lin, J.W.B., Neelin, J.D.: Influence of a stochastic moist convective parameterization on tropical climate variability. Geophys. Res. Lett. **27**, 3691–3694 (2000)
93. Lin, J.W.B., Neelin, J.D.: Considerations for stochastic convective parameterization. J. Atmos. Sci. **59**, 959–975 (2002)
94. Lin, J.W.B., Neelin, J.D.: Toward stochastic deep convective parameterization in general circulation models. Geophys. Res. Lett. **30**, 1162 (2003). https://doi.org/10.1029/2002GL016203
95. Loikith, P.C., Neelin, J.D.: Short-tailed temperature distributions over North America and implications for future changes in extremes. Geophys. Res. Lett. **42**, 8577–8585 (2015). https://doi.org/10.1002/2015GL065602
96. Lorenz, E.N.: Deterministic nonperiodic flow. J. Atmos. Sci. **20**, 130–141 (1963)
97. Lorenz, E.N.: The Essence of Chaos, 240 pp. University of Washington Press, Seattle (1995)
98. Lucarini, V., Sarno, S.: A statistical mechanical approach for the computation of the climatic response to general forcings. Nonlin. Processes Geophys. **18**, 7–28 (2011)
99. Lucarini, V., Blender, R., Herbert, C., Ragone, F., Pascale, S., Wouters, J.: Mathematical and physical ideas for climate science. Rev. Geophys. **52**, 809–859 (2014)
100. Lucarini, V., Ragone, F., Lunkeit, F.: Predicting climate change using response theory: global averages and spatial patterns. J. Stat. Phys. **166**, 1036–1064 (2017)
101. Madden, R.A., Julian, P.R.: Observations of the 40–50-day tropical oscillations – a review. Mon. Weather Rev. **122**, 814–837 (1994)
102. Majda, A., Wang, X.: Nonlinear dynamics and statistical theories for basic geophysical flows, 551 pp. Cambridge University Press, Cambridge (2006)
103. Mañé, R.: A proof of the C^1-stability conjecture. Publ. Math I.H.E.S. **66**, 161–210 (1987)
104. Mantua, N.J., Hare, S., Zhang, Y., Wallace, J.M., Francis, R.C.: A Pacific interdecadal climate oscillation with impacts on salmon production. Bull. Am. Meteorol. Soc. **78**, 1069–1079 (1997)
105. McWilliams, J.C.: Fundamentals of Geophysical Fluid Dynamics, 2nd edn., 272 pp. Cambridge University Press, Cambridge (2011)
106. Meacham, S.P.: Low-frequency variability in the wind-driven circulation. J. Phys. Oceanogr. **30**, 269–293 (2000)
107. Meehl, G.A.: Decadal climate variability and the early-2000s hiatus, vol. 13(3), pp. 1–6. In: Menemenlis, D., Sprintall, J. (eds.) US CLIVAR Variations in Understanding the Earth's Climate Warming Hiatus: Putting the Pieces Together (2015)
108. Minobe, S., Kuwano-Yoshida, A., Komori, N., Xie, S.-P., Small, R.J.: Influence of the Gulf Stream on the troposphere. Nature **452**, 206–209 (2008)
109. Mittal, A.K., Dwivedi, S., Yadav, R.S.: Probability distribution for the number of cycles between successive regime transitions for the Lorenz model. Physica D **233**, 14–20 (2007)

110. Moron, V., Vautard, R., Ghil, M.: Interannual oscillations in global sea-surface temperatures. Clim. Dyn. **14**, 545–569 (1998)
111. Nadiga, N.T., Luce, B.P.: Global bifurcation of Shilnikov type in a double-gyre ocean model. J. Phys. Oceanogr. **31**, 2669–2690 (2001)
112. Neelin, J.D., Battisti, D.S., Hirst, A.C., Jin, F.-F., Wakata, Y., Yamagata, T., Zebiak, S.: ENSO Theory. J. Geophys. Res. **104**(C7), 14261–14290 (1998)
113. Newhouse, S.: The abundance of wild hyperbolic sets and nonsmooth stable sets for diffeomorphisms. Publ. Math. I.H.E.S. **50**, 101–150 (1979)
114. NRC: Carbon Dioxide and Climate: A Scientific Assessment, Charney, J.G. et al., (eds.) National Academies Press, Washington (1979)
115. NRC: Natural Climate Variability on Decade-to-Century Time Scales, 630 pp. Martinson, D.G., Bryan, K., Ghil, M., et al., (eds.) National Academies Press, Washington (1995)
116. Ohtomo, N., et al.: Exponential characteristics of power spectral densities caused by chaotic phenomena. J. Phys. Soc. Jpn. **64**, 1104–1113 (1995)
117. Palis, J.: A global perspective for non-conservative dynamics. Ann. I.H. Poincaré **22**, 485–507 (2005)
118. Palmer, T.N.: The prediction of uncertainty in weather and climate forecasting. Rep. Prog. Phys. **63**, 71–116 (2000)
119. Palmer, T.N., Jung, T., Shutts, G.J.: Influence of a stochastic parameterization on the frequency of occurrence of North Pacific weather regimes in the ECMWF model. Geophys. Res. Lett. **32** (2005), Art. No. L23811
120. Pedlosky, J.: Geophysical Fluid Dynamics, 2nd edn., 710 pp. Springer, Berlin (1987)
121. Pedlosky, J.: Ocean Circulation Theory. Springer, New York (1996)
122. Peixoto, M.: Structural stability on two-dimensional manifolds. Topology **1**, 101–110 (1962)
123. Pierini, S.: Low-frequency variability, coherence resonance, and phase selection in a low-order model of the wind-driven ocean circulation. J. Phys. Oceanogr. **41**, 1585–1604 (2011)
124. Pierini, S.: Ensemble simulations and pullback attractors of a periodically forced double-gyre system. J. Phys. Oceanogr. **44**, 3245–3254 (2014)
125. Pierini, S., Ghil, M., Chekroun, M.D.: Exploring the pullback attractors of a low-order quasigeostrophic ocean model: the deterministic case. J. Clim. **29**, 4185–4202 (2016). https://doi.org/10.1175/JCLI-D-15-0848.1
126. Plaut, G., Ghil, M., Vautard, R.: Interannual and interdecadal variability in 335 Years of Central England temperatures. Science **268**, 710–713 (1995)
127. Poincaré, H.: Sur les équations de la dynamique et le problème des trois corps. Acta Math. **13**, 1–270 (1890)
128. Pope, V.D., Gallani, M., Rowntree, P.R., Stratton, R.A.: The impact of new physical parameterisations in the Hadley Centre climate model HadAM3. Clim. Dyn. **16**, 123–146 (2000)
129. Rasmussen, M.: Attractivity and bifurcation for nonautonomous dynamical systems. Springer, Berlin (2007)
130. Robbin, J.: A structural stability theorem. Ann. Math. **94**, 447–449 (1971)
131. Robert, R., Sommeria, J.: Statistical equilibrium states for two-dimensional flows. J. Fluid. Mech. **229**, 291–310 (1991)
132. Robinson, C.: Structural stability of C^1 diffeomorphisms. J. Differ. Equ. **22**, 28–73 (1976)
133. Ruff, T.W., Neelin, J.D.: Long tails in regional surface temperature probability distributions with implications for extremes under global warming. Geophys. Res. Lett. **39** (2012). https://doi.org/10.1029/2011GL050610
134. Salmon, R.: Lectures on Geophysical Fluid Dynamics, 378 pp. Oxford University Press, Oxford (1998)
135. Saunders, A., Ghil, M.: A Boolean delay equation model of ENSO variability. Physica D **160**, 54–78 (2001). https://doi.org/10.1029/2011GL050610
136. Schneider, S.H., Dickinson, R.E.: Climate modeling. Rev. Geophys. Space Phys. **12**, 447–493 (1974)

137. Sell, G.: Non-autonomous differential equations and dynamical systems. Trans. Am. Math. Soc. **127**, 241–283 (1967)
138. Sheremet, V.A., Ierley, G.R., Kamenkovitch, V.M.: Eigenanalysis of the two-dimensional wind-driven ocean circulation problem. J. Mar. Res. **55**, 57–92 (1997)
139. Simonnet, E., Dijkstra, H.A.: Spontaneous generation of low-frequency modes of variability in the wind-driven ocean circulation. J. Phys. Oceanogr. **32**, 1747–1762 (2002)
140. Simonnet, E., Temam, R., Wang, S., Ghil, M., Ide, K.: Successive bifurcations in a shallow-water ocean model, vol. 515, pp. 225–230. Lecture Notes in Physics, Sixteenth International Conference on Numerical Methods in Fluid Dynamics. Springer, Berlin (1995)
141. Simonnet, E., Ghil, M., Ide, K., Temam, R., Wang, S.: Low-frequency variability in shallow-water models of the wind-driven ocean circulation. Part I: Steady-state solutions. J. Phys. Oceanogr. **33**, 712–728 (2003)
142. Simonnet, E., Ghil, M., Ide, K., Temam, R., Wang, S.: Low-frequency variability in shallow-water models of the wind-driven ocean circulation. Part II: Time-dependent solutions. J. Phys. Oceanogr. **33**, 729–752 (2003)
143. Simonnet, E., Ghil, M., Dijkstra, H.A.: Homoclinic bifurcations in the quasi-geostrophic double-gyre circulation. J. Mar. Res. **63**, 931–956 (2005)
144. Simonnet, E., Dijkstra, H.A., Ghil, M.: Bifurcation analysis of ocean, atmosphere and climate models, vol. 14, pp. 187–229. Temam, R., Tribbia, J.J. (eds.) North-Holland, Amsterdam (2009)
145. Sinai, Y.: Gibbs measures in ergodic theory. Russ. Math. Surv. **27**, 21–69 (1972)
146. Smale, S.: Structurally stable systems are not dense. American J. Math. **88**(2), 491–496 (1966)
147. Smale, S.: Differentiable dynamical systems. Bull. Amer. Math. Soc. **73**, 199–206 (1967)
148. Small, R.J., DeSzoeke, S.P., Xie, S.P., O'Neill, L., Seo, H., Song, Q., Cornillon, P.: Air–sea interaction over ocean fronts and eddies. Dyn. Atmos. Oceans **45**, 274–319 (2008)
149. Soize, C.: The Fokker-Planck equation for stochastic dynamical systems and its explicit steady state solutions. World Scientific Publishing Co., Singapore (1994)
150. Speich, S., Dijkstra, H.A., Ghil, M.: Successive bifurcations in a shallow-water model applied to the wind-driven ocean circulation. Nonlinear Process. Geophys. **2**, 241–268 (1995)
151. Stainforth, D.A., et al.: Uncertainty in predictions of the climate response to rising levels of greenhouse gases. Nature **433**, 403–406 (2005)
152. Stevens, B., Zhang, Y., Ghil, M.: Stochastic effects in the representation of stratocumulus-topped mixed layers, pp. 79–90. Proceedings of ECMWF Workshop on Representation of Sub-grid Processes Using Stochastic-Dynamic Models. Shinfield Park, Reading (2005)
153. Stommel, H.: Thermohaline convection with two stable regimes of flow. Tellus **13**, 224–230 (1961)
154. Stommel, H.: The Gulf Stream: A Physical and Dynamical Description, 2nd edn., 248 pp. Cambridge University Press, London (1965)
155. Sushama, L., Ghil, M., Ide, K.: Spatio-temporal variability in a mid-latitude ocean basin subject to periodic wind forcing. Atmosphere-Ocean **45**, 227–250 (2007). https://doi.org/10.3137/ao.450404
156. Sverdrup, H.U.: Wind-driven currents in a baroclinic ocean; with application to the equatorial currents of the eastern Pacific. Proc. Natl. Acad. Sci. USA **33**, 318–326 (1947)
157. Sverdrup, H.U., Johnson, M.W., Fleming, R.H.: The Oceans: Their Physics, Chemistry and General Biology. Prentice-Hall, New York (1942). Available at http://ark.cdlib.org/ark:/13030/kt167nb66r/
158. Temam, R.: Infinite-Dimensional Dynamical Systems in Mechanics and Physics, 2nd edn., 648 pp. Springer, New York (1997)
159. Thompson, P.D.: Numerical Weather Analysis and Prediction, 170 pp. Macmillan, New York (1961)
160. Trefethen, L.N., Trefethen, A., Reddy, S.C., Driscoll, T.A.: Hydrodynamic stability without eigenvalues. Science **261**, 578–584 (1993)
161. Tucker, W.: Lorenz attractor exists. C. R. Acad. Sci. Paris **328**(12), 1197–1202 (1999)

162. Tziperman, E., Stone, L., Cane, M., Jarosh, H.: El Niño chaos: overlapping of resonances between the seasonal cycle and the Pacific ocean-atmosphere oscillator. Science **264**, 72–74 (1994)
163. Tziperman, E., Cane, M.A., Zebiak, S.E.: Irregularity and locking to the seasonal cycle in an ENSO prediction model as explained by the quasi-periodicity route to chaos. J. Atmos. Sci. **50**, 293–306 (1995)
164. Vallis, G.: Atmospheric and Oceanic Fluid Dynamics, 745 pp. Cambridge University Press, Cambridge (2006)
165. Vannitsem, S.: Dynamics and predictability of a low-order wind-driven ocean–atmosphere coupled model. Clim. Dyn. **42**, 1981–1998 (2014)
166. Vannitsem, S., Ghil, M.: Evidence of coupling in ocean–atmosphere dynamics over the North Atlantic. Geophys. Res. Lett. **44**, 2016–2026 (2017). https://doi.org/10.1002/2016GL072229
167. Vannitsem, S., Demaeyer, J., De Cruz, L., Ghil, M.: A 24-variable low-order coupled ocean-atmosphere model: OA-QG-WS v2. Geosci. Model Dev. **7**, 649–662 (2014)
168. Vannitsem, S., Demaeyer, J., De Cruz, L., Ghil, M.: Low-frequency variability and heat transport in a low-order nonlinear coupled ocean-atmosphere model. Physica D **309**, 71–85 (2015). https://doi.org/10.1016/j.physd.2015.07.006
169. Weeks, E.R., Tian, Y., Urbach, J.S., Ide, K., Swinney, H.L., Ghil, M.: Transitions between blocked and zonal flows in a rotating annulus with topography. Science **278**, 1598–1601 (1997)
170. Williams, R.F.: The structure of Lorenz attractors. Publ. Math. I.H.E.S. **50**, 73–99 (1979)
171. Wunsch, C.: The interpretation of short climate records, with comments on the North Atlantic and Southern Oscillations. Bull. Am. Meteorol. Soc. **80**, 245–255 (1999)
172. Young, L.S.: What are SRB measures, and which dynamical systems have them? J. Stat. Phys. **108**, 733–754 (2002)

Parameter Determination for Energy Balance Models with Memory

Piermarco Cannarsa, Martina Malfitana, and Patrick Martinez

Abstract In this paper, we study two Energy Balance Models with Memory arising in climate dynamics, which consist in a 1D degenerate nonlinear parabolic equation involving a memory term, and possibly a set-valued reaction term (of Sellers type and of Budyko type, in the usual terminology). We provide existence and regularity results, and obtain uniqueness and stability estimates that are useful for the determination of the insolation function in Sellers' model with memory.

Keywords Parameter determination · Energy balance models · Degenerate Parabolic equations

1 Introduction

1.1 Energy Balance Models and the Problems We Consider

We are interested in a problem arising in climate dynamics, coming more specifically from the classical Energy Balance models introduced independently by the climatologists Budyko [4] and Sellers [39], and written from a mathematical point of view respectively by Held and Suarez [25] and Ghil [22]. These models, which describe the evolution of temperature as the effect of the balance between the amount of energy received from the Sun and radiated from the Earth, were developed in order to understand the past and future climate and its sensitivity to some

P. Cannarsa (✉) · M. Malfitana
Dipartimento di Matematica, Università di Roma "Tor Vergata", Roma, Italy
e-mail: cannarsa@mat.uniroma2.it

P. Martinez
Institut de Mathématiques de Toulouse, UMR 5219, Université de Toulouse, CNRS UPS IMT, Toulouse, France
e-mail: patrick.martinez@math.univ-toulouse.fr

© Springer Nature Switzerland AG 2020
P. Cannarsa et al. (eds.), *Mathematical Approach to Climate Change and Its Impacts*, Springer INdAM Series 38,
https://doi.org/10.1007/978-3-030-38669-6_2

relevant parameters on large time scales (centuries). After averaging the surface temperature over longitude, they take the form of the following one-dimensional nonlinear parabolic equation with degenerate diffusion:

$$u_t - (\rho_0(1 - x^2)u_x)_x = R_a - R_e$$

where

- $u(t, x)$ is the surface temperature averaged over longitude,
- the space variable $x = \sin \phi \in (-1, 1)$ (here ϕ denotes the latitude),
- R_a represents the fraction of solar energy absorbed by the Earth,
- R_e represents the energy emitted by the Earth,
- ρ_0 is a positive parameter.

A crucial role in the analysis will be played by the absorbed energy R_a, which is a fraction of the incoming solar flux $Q(t, x)$, that is,

$$R_a = Q(t, x)\,\beta,$$

where β is the coalbedo function. Additionally, as is customary in seasonally averaged models, we will assume that

$$Q(t, x) = r(t)q(x),$$

where r is positive and q is the so-called "insolation function".

It was noted (see Bhattacharya et al. [2]) that, in order to take into account the long response times that cryosphere exhibits (for instance, the expansion or retreat of huge continental ice sheets occurs with response times of thousands of years), it is useful to let the coalbedo function depend not only on u, but also on the history function, which can be represented by the integral term

$$H(t, x, u) := \int_{-\tau}^{0} k(s, x)u(t + s, x)ds \qquad \forall t > 0, x \in I,$$

where k is the memory kernel (and $\tau \sim 10^4$ years, in real problems). As in Roques et al. [38], we will assume a nonlinear response to memory in the form

$$f(H(t, x, u)).$$

Hence, we are interested in the following Energy Balance Model with Memory (EBBM) problem, set in the space domain $I := (-1, 1)$:

$$\begin{cases} u_t - (\rho_0(1 - x^2)u_x)_x = Q(t, x)\beta(u) + f(H(t, x, u)) - R_e(u), & t > 0, x \in I, \\ \rho_0(1 - x^2)u_x(t, x) = 0, & t > 0, x \in \partial I \\ u(s, x) = u_0(s, x), & s \in [-\tau, 0]. \end{cases}$$

Concerning the function β, we will assume, as it is classical for such problems, that

- either β is positive and at least Lipschitz continuous (the classical assumption for Sellers type models),
- or β is positive, monotone and discontinuous (the classical assumption for Budyko type models).

1.2 Relation to Literature and Presentation of Our Main Results

The mathematical analysis of quasilinear EBMM problems of the form

$$\partial_t u - \mathrm{div}\, (\rho(x)|\nabla u|^{p-2}\nabla u) = f(t, x, u, H(t, x, u))$$

has been the subject of many deep works for a long time. Questions such as well-posedness, uniqueness, asymptotic behavior, existence of periodic solutions, bifurcation, free boundary, numerical approximation were investigated for:

- 1-D models without memory by Ghil in the seminal paper [22] (and see also [3, 44]),
- 0-D models in Fraedrich [20, 21],
- 1-D models with memory in Bhattacharya et al. [2] and Diaz [12, 13],
- 2-D models (on a manifold without boundary, typically representing the Earth's surface) in Hetzer [26], Diaz and Tello [16], Diaz and Hetzer [15], Hetzer [27], Diaz [14], Diaz et al. [17], and Hetzer [28].

In this paper, we are interested in the following inverse problem: is it possible to recover the insolation function (which is a part of the incoming solar flux in $Q(t, x)$) from measurements of the solution, for our EBBM model? Our motivation comes from the fact that, with suitable tuning of their parameters, EBMs have shown to mimic the observed zonal temperatures for the observed present climate [33], and can be used to estimate the temporal response patterns to various forcing scenarios, which is of interest in particular in the detection of climate change. Unfortunately, in practice, the model coefficients cannot be measured directly, but are quantified through the measures of the solution [38]. Hence, results proving that measuring the solution in some specific (small) part of the space and time domain is sufficient to recover a specified coefficient are of practical interest.

Several earlier papers are related to this question, in particular the ones that we recall below.

- In Tort and Vancostenoble [42], the question of determining the insolation function was studied for a 1D Sellers type model without memory, combining:
    - the method introduced by Imanuvilov and Yamamoto in the seminal paper [29] (based on the use of Carleman estimates to obtain stability results for the determination of source terms for parabolic equations),
    - the Carleman estimates from Cannarsa et al. [7] for degenerate parabolic equations,
    - suitable maximum principles to deal with nonlinear terms.

  In the same paper, the authors proved stability estimates measuring the solution on an open subset of the space domain. Similar questions were studied in Martinez et al. [32] on manifolds without boundary.
- In Roques et al. [38], the question of determining the insolation function was studied for a 1D Sellers type model including memory effects, but for a nondegenerate diffusion. These authors extended a method due to Roques and Cristofol [11, 37] which, based on analyticity, allows for measurements only at a single point x_0 (under a rather strong assumption on the kernel appearing in the history function).

In this paper we study, first, the 1D Sellers type problem with degenerate diffusion and memory effects. More precisely, we prove regularity results and use them to study the determination of the insolation function, obtaining

- a uniqueness result, under pointwise observation,
- a Lipschitz stability result, under localized observation,

in the spirit of the above mentioned references. Then, we address 1D Budyko type problems with degenerate diffusion and memory effects, for which we obtain precise existence results as in Diaz and Hetzer [15]. For this, we need to regularize the coalbedo and use the existence results obtained in the first part of the paper.

Let us note that our existence results for Sellers and Budyko type problems can be regarded as a consequence of the ones by Diaz and Hetzer [15] for manifolds. However, here we give a direct proof of such results in zonally averaged 1D settings. For this reason, we need to use the properties of degenerate diffusion operators.

Finally, to give a more complete overview of the literature on these questions, let us also mention the papers by:

- Pandolfi [34], for a similar question but on a different equation (the history function depending on the second-order derivative in space),
- Guerrero and Imanuvilov [24], that proves that null controllability does *not* hold for the linear heat equation perturbed by $\int_0^t u$ (hence with a memory term which takes into account all the history from time 0 to t),
- Tao and Gao [41], that gives positive null controllability results for a similar heat equation under additional assumptions on the kernel appearing in the history

function (notice however that these assumptions are incompatible with our settings as they would force the kernel to depend also t and to uniformly vanish at some time T, which is unnatural in climate modelling),
- Floridia et al. [10, 18, 19], that studied the controllability of Sellers type models using bilinear controls, hence when the control is a multiplicative parameter.

2 Mathematical Assumptions for These Climate Models

We are interested in a class of EBMM:

$$\begin{cases} u_t - (\rho(x)u_x)_x = R_a(t, x, u, H) - R_e(t, x, u, H)), & t > 0, x \in I, \\ \rho(x)u_x = 0, & x \in \partial I, \\ u(s, x) = u_0(s, x), & s \in [-\tau, 0], x \in I, \end{cases} \quad (1)$$

where $I = (-1, 1)$. We are going to precise our assumptions concerning Budyko type problems and Sellers ones.

2.1 Budyko Type Models with Memory

We make the following assumptions:

- concerning the diffusion coefficient: we assume that there exists $\rho_0 > 0$ such that

$$\forall x \in (-1, 1), \quad \rho(x) := \rho_0(1 - x^2); \quad (2)$$

- concerning R_a: we assume that

$$R_a(t, x, u, H) = Q(t, x)\beta(u) + f(H(t, x, u)), \quad (3)$$

where

- $Q(t, x)$ is the incoming solar flux; we assume that $Q(t, x) = r(t)q(x)$, where q, the insolation function, and r are such that:

$$\begin{cases} q \in L^\infty(I), \\ r \in C^1([0, +\infty)) \text{ and } r, r' \in L^\infty([0, +\infty)); \\ \exists r_1 > 0 \text{ s.t. } \forall t \geq 0, \quad r(t) \geq r_1; \end{cases} \quad (4)$$

- β is the classical Budyko type coalbedo function: it is an highly variable quantity which depends on many local factors such as the cloud cover and the composition of the Earth's atmosphere, moreover it is used as an indicator for ice and snow cover; usually it is considered roughly constant for temperatures far enough from the ice-line, that is a circle of constant latitude that separates the polar ice caps from the lower ice-free latitudes; the classical Budyko type coalbedo is:

$$\beta(u) = \begin{cases} a_i, & u < \bar{u}, \\ [a_i, a_f], & u = \bar{u}, \\ a_f, & u > \bar{u}, \end{cases} \tag{5}$$

where $a_i < a_f$ (and the threshold temperature $\bar{u} := -10°$);
– H is the history function; it is assumed to be given by

$$H(t, x, u) = \int_{-\tau}^{0} k(s, x) u(t + s, x) \, ds \tag{6}$$

where the kernel k is such that:

$$k \in C^1([-\tau, 0] \times [-1, 1]; (-\infty, +\infty)); \tag{7}$$

– f: the nonlinearity that describes the memory effects; we assume that $f : (-\infty, +\infty) \to (-\infty, +\infty)$ is C^1 and such that

$$\begin{cases} f, f' \in L^{\infty}(-\infty, +\infty) \\ f, f' \text{ are } L - \text{Lipschitz}; \end{cases} \tag{8}$$

• concerning R_e: the classical Budyko type assumption is

$$R_e(t, x, u, H) = a + bu, \tag{9}$$

where a, b are constants;
• the initial condition: since we define H over a past temperature, the initial condition in such models has to be of the form

$$u(s, x) = u_0(s, x) \quad \forall s \in [-\tau, 0], \quad x \in I \tag{10}$$

for some $u_0(s, x)$ defined on $[-\tau, 0] \times I$, for which we will precise our assumptions in our different results.

Sometimes we will only add positivity assumptions on q and r; these assumptions are natural with respect to the model, but only useful in the inverse problems results.

2.2 Sellers Type Models with Memory

The differences concern the assumptions on the coalbedo and on the emitted energy:

- β: in Sellers type models, we assume that

$$\beta \in C^2(-\infty, +\infty), \quad \beta, \beta', \beta'' \in L^\infty(-\infty, +\infty) \tag{11}$$

(typically, β is C^2 and takes values between the lower value for the coalbedo a_i and higher value a_f (even if there is a sharp transition between these two values around the threshold temperature $\bar{u}$)).

- R_e is assumed to follow a Stefan-Boltzmann type law (assuming that the Earth radiates as a black body):

$$R_e = \varepsilon(u)|u|^3 u, \tag{12}$$

where the function ε represents the emissivity; we assume that

$$\begin{cases} \varepsilon \in C^1(-\infty, +\infty) \text{ and } \varepsilon, \varepsilon' \in L^\infty(-\infty, +\infty), \\ \exists \varepsilon_1 > 0, \text{ s.t. for all } u, \quad \varepsilon(u) \geq \varepsilon_1 > 0. \end{cases} \tag{13}$$

We also refer the reader to the table 10.1 of Ghil and Childress [23], that allows one to easily see and compare the assumptions of the two problems.

2.3 Plan of the Paper

- Section 3 contains the statement of our results concerning Sellers type models:
  - concerning well-posedness questions: see Theorem 2 in Sect. 3.3;
  - concerning inverse problems questions:

    Theorem 3: uniqueness of the insolation function under pointwise measurements (in Sect. 3.4.1),

    Theorem 4: Lipschitz stability under localized measurements (in Sect. 3.4.2);

- Section 4 contains the statement of our well-posedness result concerning Budyko type models, see Theorem 5;
- Section 5 is devoted to mention some open questions;
- Section 6 contains the proof of Theorem 2;
- Section 7 contains the proof of Theorem 3;

- Section 8 contains the proof of Theorem 4;
- Section 9 contains the proof of Theorem 5.

3   Main Results for the Sellers Type Model

First we show the local and global existence of a regular solution to the following problem:

$$
\begin{cases}
u_t - (\rho(x)u_x)_x = r(t)q(x)\beta(u) - \varepsilon(u)|u|^3 u + f(H), & t > 0, x \in I, \\
\rho(x)u_x = 0, & t > 0, x \in \partial I, \\
u(s, x) = u_0(s, x), & s \in [-\tau, 0], x \in I,
\end{cases}
\tag{14}
$$

In this section, we assume (2), (4), (6)–(8), (10)–(13).

In the following, we recast (14) into a semilinear evolution equation governed by an analytic semigroup.

3.1   Functional Framework

Since the diffusion coefficient has a degeneracy at the boundary, it is necessary to introduce the weighted Sobolev space V below in order to deal the well-posedness of problem (1). To know more about this functional framework for one-dimensional degenerate parabolic equations, the reader may also refer to [5, 6, 8, 42].

$$
V := \{w \in L^2(I) : w \in AC_{loc}(I), \sqrt{\rho}w_x \in L^2(I)\}
$$

endowed with the inner product

$$
(u, v)_V := (u, v)_{L^2(I)} + (\sqrt{\rho}u_x, \sqrt{\rho}v_x)_{L^2(I)} \quad \forall u, v \in V
$$

and then with the associated norm

$$
\|u\|_V := \sqrt{(u, u)_V} = \|u\|_{L^2(I)} + \|\sqrt{\rho}u_x\|_{L^2(I)} \quad \forall u \in V.
$$

We recall that $\rho(x) = \rho_0(1 - x^2)$ for all $x \in I$ by definition. Let us remark that $(V, (\cdot, \cdot)_V)$ is a Hilbert space and that $V \subset H^1_{loc}(I)$ and $V \subset L^2(I) \subset V^*$. Moreover,

- the space $C_0^\infty(I)$ is dense in V, in particular V is dense in $L^2(I)$ [5, Lemma 2.6];

- for all $p \in [1, +\infty)$, the inclusion

$$V \hookrightarrow L^p(I) \tag{15}$$

holds and is continuous; moreover, the inclusion $V \hookrightarrow L^2(I)$ is compact [13, Lemma 1].

In order to obtain our semilinear evolution equation let us define the operator $A : D(A) \subset L^2(I) \rightarrow L^2(I)$ in the following way:

$$\begin{cases} D(A) := \{u \in V : \rho u_x \in H^1(I)\} \\ Au := (\rho u_x)_x \quad u \in D(A) \end{cases} \tag{16}$$

(Note that the boundary condition appearing in (1) is contained in the definition of the unbounded operator A given in (16): indeed, if $u \in D(A)$, then $\rho u_x \in H^1(I)$, hence $\rho u_x \in C^0(\bar{I})$, which implies that $\rho u_x \rightarrow L$ as $x \rightarrow 1^-$; but if $L \neq 0$, then $\sqrt{\rho} u_x \notin L^2(I)$, therefore $L = 0$ and $(\rho u_x)(1) = 0$. And the case $x = -1$ is analogous.)

We denote $\mathcal{L}(L^2(I))$ the space of linear continuous applications from $L^2(I)$ into itself, endowed with the natural norm $||| \cdot |||_{\mathcal{L}(L^2(I))}$. We recall the following

Theorem 1 ([1, 5]) *(A,D(A)) is a self-adjoint operator and it is the infinitesimal generator of an analytic and compact semigroup $\{e^{tA}\}_{t\geq0}$ in $L^2(I)$ that satisfies*

$$|||e^{tA}|||_{\mathcal{L}(L^2(I))} \leq 1.$$

(We give elements of its proof in Sect. 6.1). Finally, we recall also the following

Proposition 1 ([31, Proposition 2.1]) *The real interpolation space constructed by the trace method $[D(A), L^2(I)]_{\frac{1}{2}}$ is the space V.*

3.2 The Concept of Mild Solution for the Sellers Type Model (14)

Consider the problem (1) of the Sellers type. In order to recast it into an abstract form, we introduce the following notations:

- to manage the nonlinear term, we consider the following function

$$G : [0, T] \times V \rightarrow L^2(I), \quad G(t, u)(x) = Q(t, x)\beta(u(x)) - \varepsilon(u(x))|u(x)|^3 u(x), \tag{17}$$

where we recall that $Q(t, x) = r(t)q(x)$; (note that since $V \hookrightarrow L^p(I)$ for all $p \geq 1$, it is clear that $G([0, T] \times V) \subset L^2(I)$);

- to manage the shifted memory term:

  - given $u \in C([-\tau, T]; L^2(I))$, given $t \in [0, T]$, we consider the right translation $u^{(t)} \in C([-\tau, 0]; L^2(I))$ by the formula

$$u^{(t)} : [-\tau, 0] \to L^2(I), \quad u^{(t)}(s) := u(t + s), \tag{18}$$

  - and we define the following function

$$F : C([-\tau, 0]; L^2(I)) \to L^2(I), \quad F(v)(x) = f \left(\int_{-\tau}^{0} k(s, x) (v(s)(x)) \, d\sigma \right), \tag{19}$$

in such a way that the memory term can be written $F(u^{(t)})$.

And then, given $T > 0$, (14) on $[0, T]$ can be recast into:

$$\begin{cases} \dot{u}(t) = Au(t) + G(t, u) + F(u^{(t)}) \ t \in [0, T] \\ u(s) = u_0(s) \hspace{3.5cm} s \in [-\tau, 0]. \end{cases} \tag{20}$$

Before defining the concept of mild solution for (20), we precise the concept of mild solution for the following linear nonhomogeneous problem

$$\begin{cases} \dot{u}(t) = Au(t) + g(t) \ t \in [0, T] \\ u(0) = u_0. \end{cases} \tag{21}$$

We consider the following

Definition 1 Let $g \in L^2(0, T; L^2(I))$ and let $u_0 \in L^2(I)$. The function $u \in C([0, T]; L^2(I))$ defined by

$$\forall t \in [0, T], \quad u(t) = e^{tA}u_0 + \int_0^t e^{(t-s)A} g(s) ds \tag{22}$$

is called the mild solution of (21).

We recall that u defined by (22) has the following additional regularity:

$$u \in H^1(0, T; L^2(I)) \cap L^2(0, T; D(A)).$$

Now we are ready to define the concept of mild solution for (20):

Definition 2 Given $u_0 \in C([-\tau, 0]; V)$, a function

$$u \in H^1(0, T; L^2(I)) \cap L^2(0, T; D(A)) \cap C([-\tau, T]; V)$$

is called a mild solution of (20) on $[0, T]$ if

- $u(s) = u_0(s)$ for all $s \in [-\tau, 0]$;
- for all $t \in [0, T]$, we have

$$u(t) = e^{tA}u_0(0) + \int_0^t e^{(t-s)A}\big(G(s, u) + F(u^{(s)})\big)\, ds. \qquad (23)$$

3.3 Global Existence and Uniqueness Result for the Sellers Model (20)

Now we are ready to prove the global existence result of the integrodifferential problem.

Theorem 2 *Consider u_0 such that*

$$u_0 \in C([-\tau, 0]; V) \quad and \quad u_0(0) \in D(A) \cap L^\infty(I).$$

Then, for all $T > 0$, the problem (20) has a unique mild solution u on $[0, T]$.

(Note that

- existence and uniqueness of a global regular solution to (1) without the memory term has been proved in [42];
- the local existence of our model without the boundary degeneracy has been studied in [38];
- the global existence of a similar 2D-model with memory (hence on a manifold but without the boundary degeneracy), has been investigated in [15].)

3.4 Inverse Problem Results: Determination of the Insolation Function

Here we prove that the insolation function $q(x)$ can be determined in the whole space domain I by using only local information about the temperature.

To achieve this goal, we add the following extra assumptions, as in [38]: the very recent past temperatures are not taken into account in the history function:

$$\exists \delta > 0 \text{ s.t. } k(s, \cdot) \equiv 0 \quad \forall s \in [-\delta, 0] \qquad (24)$$

where $\delta < \tau$. (We will discuss about this assumption in Sect. 5.)

Hence, we have the following situation: consider two insolation functions q and $\tilde{q}$, two initial conditions u_0 and $\tilde{u}_0$, and the associated solutions: u satisfying (14) and $\tilde{u}$ satisfying

$$
\begin{cases}
\tilde{u}_t - (\rho(x)\tilde{u}_x)_x = r(t)\tilde{q}(x)\beta(\tilde{u}) - \varepsilon(\tilde{u})|\tilde{u}|^3\tilde{u} + f(\tilde{H}), & t > 0, x \in I, \\
\rho(x)\tilde{u}_x = 0, & x \in \partial I, \\
\tilde{u}(s, x) = \tilde{u}_0(s, x), & s \in [-\tau, 0], x \in I,
\end{cases}
$$

$$(25)$$

where we denote

$$
\tilde{H} := H(t, x, \tilde{u}) = \int_{-\tau}^{-\delta} k(s, x)\tilde{u}(t + s, x)\, ds.
$$

In the following, we state two inverse problems results, according to different assumptions on the control region.

3.4.1 Pointwise Observation and Uniqueness Result

Let us choose suitable regularity assumptions on the initial conditions and on the insolation functions, in order to have sufficient regularity on the time derivative of the associated solutions: we consider

- the set of admissible initial conditions: we consider

$$
\mathcal{U}^{(pt)} = C^{1,2}([-\tau, 0] \times [-1, 1]),
$$

$$(26)$$

- and the set of admissible coefficients: we consider

$$
\mathcal{Q}^{(pt)} := \{q \text{ is Lipschitz-continuous and piecewise analytic on } I\},
$$

$$(27)$$

where we recall the following

Definition 3 A continuous function ψ is called *piecewise analytic* if there exist $n \geq 1$ and an increasing sequence $(p_j)_{1 \leq j \leq n}$ such that $p_1 = -1$, $p_n = 1$, and

$$
\psi(x) = \sum_{j=1}^{n-1} \chi_{[p_j, p_{j+1})}(x)\varphi_j(x) \quad \forall x \in I,
$$

where φ_j are analytic functions defined on the intervals $[p_j, p_{j+1}]$ and $\chi_{[p_j, p_{j+1})}$ is the characteristic function of the interval $[p_j, p_{j+1})$ for $j = 1, \ldots, n-1$.

Then we prove the following uniqueness result:

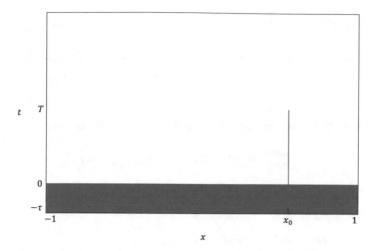

Fig. 1 Space-time measurement region which can lead to unique coefficient determination

Theorem 3 *Consider*

- *two insolation functions $q, \tilde{q} \in \mathcal{Q}^{(pt)}$ (defined in (27))*
- *an initial condition $u_0 = \tilde{u}_0 \in \mathcal{U}^{(pt)}$ (defined in (26))*

and let u be the solution of (14) and $\tilde{u}$ the solution of (25).
Assume that

- *the memory kernel satisfies (24),*
- *r and β are positive,*
- *there exists $x_0 \in I$ and $T > 0$ such that*

$$\forall t \in (0, T), \quad \begin{cases} u(t, x_0) = \tilde{u}(t, x_0), \\ u_x(t, x_0) = \tilde{u}_x(t, x_0). \end{cases} \tag{28}$$

Then $q \equiv \tilde{q}$ on I.

This result means that the insolation function $q(x)$ is uniquely determined on I by any measurement of u and u_x at a single point x_0 during the time period $(0, T)$. Theorem 3 is a natural extension of [38] to the degenerate problem. The region that is sufficient to prove the uniqueness of q is showed in Fig. 1. (Note that for the proof of Theorem 3 we will assume that $T < \delta$.)

3.4.2 Localized Observation and Stability Result

Let us choose suitable regularity assumptions on the initial conditions and on the insolation functions, in order to have sufficient regularity on the time derivative of the associated solutions: we consider

- the set of admissible initial conditions: given $M > 0$, we consider $\mathcal{U}_M^{(loc)}$:

$$\mathcal{U}_M^{(loc)} := \{u_0 \in C([-\tau, 0]; V \cap L^\infty(-1, 1)), u_0(0) \in D(A), Au_0(0) \in L^\infty(I),$$
$$\sup_{t \in [-\tau, 0]} (\|u_0(t)\|_V + \|u_0(t)\|_{L^\infty}) + \|Au_0(0)\|_{L^\infty(I)} \leq M\},$$
(29)

- and the set of admissible coefficients: given $M' > 0$, we consider

$$\mathcal{Q}_{M'}^{(loc)} := \{q \in L^\infty(I) : \|q\|_{L^\infty(I)} \leq M'\}.$$
(30)

Now we are ready to state our Lipschitz stability result:

Theorem 4 *Assume that*

- *the memory kernel satisfies (24),*
- *r and β are positive.*

Consider

- *$0 < T' < \delta$,*
- *$t_0 \in [0, T')$, $T > T'$,*
- *$M, M' > 0$.*

Then there exists $C(t_0, T', T, M, M') > 0$ such that, for all $u_0, \tilde{u}_0 \in \mathcal{U}_M^{(loc)}$ (defined in (29)), for all $q, \tilde{q} \in \mathcal{Q}_{M'}^{(loc)}$ (defined in (30)), the solution u of (14) and the solution $\tilde{u}$ of (25) satisfy

$$\|q - \tilde{q}\|_{L^2(I)}^2 \leq C \left(\|u(T') - \tilde{u}(T')\|_{D(A)}^2 \right.$$
$$\left. + \|u_t - \tilde{u}_t\|_{L^2((t_0, T) \times (a, b))}^2 + \|u_0 - \tilde{u}_0\|_{C([-\tau, 0]; V)}^2 \right).$$
(31)

Theorem 4 is a natural extension of [42]. The region that is sufficient to prove the Lipschitz stability is showed in Fig. 2. (Note that for the proof of Theorem 4 we will assume that $T < \delta$.)

4   Main Result for the Budyko Type Model

Now we treat the global existence of regular solutions for the Budyko model. In a classical way (see, e.g. Diaz [13]), we study the set valued problem

- first regularizing the coalbedo, hence transforming the Budyko type problem into a Sellers one, for which we have a (unique) regular solution,
- and then passing to the limit with respect to the regularization parameter.

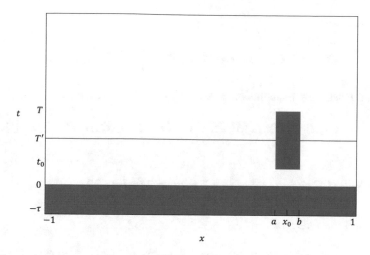

Fig. 2 Space-time measurement region which can lead to Lipschitz stability estimates

Since β is the graph given in (5), the Budyko type problem has to be understood as the following differential inclusion problem:

$$\begin{cases} u_t - (\rho(x)u_x)_x \in r(t)q(x)\beta(u) - (a+bu) + f(H(u)), & t > 0, x \in I, \\ \rho(x)u_x = 0, & x = \pm 1, \\ u(s,x) = u_0(s,x), & s \in [-\tau, 0], x \subset I. \end{cases}$$

$$(32)$$

In this section, we assume (2)–(10).

4.1 The Notion of Mild Solutions for the Budyko Model (32)

Let us define a mild solution for this kind of problem.

Definition 4 Given $u_0 \in C([-\tau, 0); V)$, a function

$$u \in H^1(0, T; L^2(I)) \cap L^2(0, T; D(A)) \cap C([-\tau, T]; V)$$

is called a *mild solution* of (32) on $[-\tau, T]$ iff

- $u(s) = u_0(s)$ for all $s \in [-\tau, 0]$;
- there exists $g \in L^2([0, T]; L^2(I))$ such that

– *u* satisfies

$$\forall t \in [0, T], \quad u(t) = e^{tA}u_0(0) + \int_0^t e^{(t-s)A}g(s)\, ds, \tag{33}$$

– and *g* satisfies the inclusion: a.e. $(t, x) \in (0, T) \times I$, we have

$$g(t, x) \in r(t)q(x)\beta(u(t, x)) - (a + bu(t, x)) + f(H(t, x, u)).$$

4.2 Global Existence for the Budyko Model (32)

Theorem 5 *Assume that*

$$u_0 \in C([-\tau, 0], V) \quad and \quad u_0(0) \in D(A) \cap L^\infty(I).$$

Then (32) has a mild solution u, which is global in time (i.e. defined in $[0, +\infty)$ and mild on $[-\tau, T]$ for all $T > 0$).

5 Open Questions

Let us mention some open questions related to this work.

- Concerning the Sellers type models and the inverse problems results given in Theorems 3 and 4: the assumption on the support of the kernel *k* is crucial, since it allows us to easily get rid of the memory term, and we do not know what can be done without this assumption; hence

  - it would be interesting to weaken this support assumption; it seems reasonable to think that uniqueness and stability results could be obtained in a more general context, and on the other hand, memory terms can sometimes generate problems (see, in particular [24]);
  - even under the support assumption: from a numerical point of view, it would be interesting to weaken some assumptions (in particular on *T*, since our proof is based on $T < \delta$) in order to have better estimates even if δ is small.

- Concerning the Budyko type models: a solution is obtained by regularization and passage to the limit; (note that, at least for an EBM without memory term, uniqueness of the solution depends on the initial condition, see Diaz [13]); several questions are mathematically challenging, in particular to obtain inverse problems results in that setting; one way could be to obtain suitable estimates from the regularized problem, succeeding in avoiding to use the C^1 norm of the

regularized coalbedo (unfortunately, the C^1 norm of the regularized coalbedo appears in our estimates in the Sellers model).
- It would also be interesting to obtain inverse problems results for other problems, in particular of the quasilinear type mentioned in the beginning of Sect. 1.2.
- Finally, we hope that this analysis could be useful to study the proximity of the climate system to one of the so-called "tipping points", see in particular [30, 45], but we do not address such a question in this paper.

6 Proof of Theorem 2

6.1 Elements for Theorem 1

To prove that $(A, D(A))$ is self-adjoint and it generates a strongly continuous semigroup of contractions in $L^2(I)$, one can refer to Theorem 2.8 in [5] (where in our case $b \equiv 0$).

Moreover,

$$\int_0^x \frac{ds}{\rho_0(1 - s^2)} = \frac{1}{2\rho_0}(\log(1 + x) - \log(1 - x))$$

which implies that $\int_0^x \frac{ds}{\rho(s)} \in L^1(I)$, and then one can show that the semigroup generated by A is compact (see Theorem 3.3 in [5]).

For the analyticity of the semigroup see Theorem 2.12 in [1] or Theorem 3.6.1 in [40].

6.2 Proof of Theorem 2: Local Existence of Mild Solutions

In this section, we prove the following

Proposition 2 *Consider u_0 such that*

$$u_0 \in C([-\tau, 0]; V) \quad and \quad u_0(0) \in D(A).$$

Then there exists $t^* > 0$ such that the problem (20) has a unique mild solution on $[0, t^*]$.

6.2.1 The Functional Setting and Main Tools

It will be more practical to have strictly dissipative operators, so let us consider $\tilde{A} := A - I$, that satisfies:

$$\begin{cases} D(\tilde{A}) = D(A) \\ \tilde{A}u := (\rho u_x)_x - u, \ u \in D(\tilde{A}) : \end{cases}$$

integrating by parts:

$$\forall u \in D(\tilde{A}), \quad (\tilde{A}u, u)_{L^2(I)} = -\int_I (\rho u_x^2 + u^2)dx \leq -\|u\|^2_{L^2(I)},$$

hence $\tilde{A}$ is strictly dissipative. We will use the following estimates: using Pazy [35] (Theorem 6.13, p. 74) with $\alpha = 1/2$ and $\alpha = 3/4$, there exists $c > 0$ such that

$$\forall t > 0, \quad |||(-\tilde{A})^{1/2}e^{t\tilde{A}}|||_{\mathcal{L}(L^2(I))} \leq \frac{c}{\sqrt{t}}, \tag{34}$$

and

$$\forall t > 0, \quad |||(-\tilde{A})^{3/4}e^{t\tilde{A}}|||_{\mathcal{L}(L^2(I))} \leq \frac{c}{t^{3/4}}. \tag{35}$$

Now, consider also $\tilde{G}(t, u) := u + G(t, u)$ so that, adding and subtracting u to the equation, the problem (20) is equivalent to

$$\begin{cases} \dot{u}(t) = \tilde{A}u(t) + \tilde{G}(t, u(t)) + F(u^{(t)}) \\ u(0) = u_0(0). \end{cases} \tag{36}$$

By the definitions above, we know that a mild solution of this problem is a function such that

$$u(t) = \begin{cases} e^{t\tilde{A}}u_0(0) + \int_0^t e^{(t-s)\tilde{A}}\left[\tilde{G}(s, u(s)) + F(u^{(s)})\right] ds, \ t > 0 \\ u_0(t), \hspace{5cm} t \in [-\tau, 0] \end{cases}$$

Then we consider the suitable functional setting:

- the space of functions

$$\mathcal{X}_R := \left\{ v \in C([-\tau, t^*]; V) \ \Big| \ \begin{cases} \|v(t)\|_V \leq R \ \forall t \in [-\tau, t^*], \\ v(t) = u_0(t) \ \forall t \in [-\tau, 0], \end{cases} \right\},$$

- and the associated application

$$\Gamma : \mathcal{X}_R \subset C([-\tau, t^*]; V) \to C([-\tau, t^*]; V)$$

defined by

$$\Gamma(u)(t) := \begin{cases} e^{t\tilde{A}}u_0(0) + \int_0^t e^{(t-s)\tilde{A}} \left[\tilde{G}(s, u(s)) + F(u^{(s)}) \right] ds, \ t \in [0, t^*] \\ u_0(t), \hspace{6cm} t \in [-\tau, 0]. \end{cases}$$

In the following, we prove that Γ maps $\mathcal{X}_R$ into itself and is a contraction if the parameters R and t^* are well-chosen. Then the Banach–Caccioppoli fixed point theorem will tell us that Γ has a fixed point, and this fixed point will be a mild solution of our problem.

We will use the following properties of the function G:

Lemma 1 ([42, Lemma 3.4]) *Take $R > 0$. Then G is well defined on $[0, T] \times V$ with values into $L^2(I)$. Moreover, we have the following estimates:*

1. there exists $C_R > 0$ such that

$$\begin{cases} \|G(t, u)\|_{L^2(I)} \le C_R(1 + \|u\|_V), \\ \forall t \in [0, T], \forall u \in V \ s.t. \ \|u\|_V \le R, \end{cases} \tag{37}$$

and

$$\begin{cases} \|G(t, u) - G(t, v)\|_{L^2(I)} \le C_R \|u - v\|_V, \\ \forall t \in [0, T], \forall u, v \in V \ s.t. \ \|u\|_V, \|v\|_V \le R; \end{cases} \tag{38}$$

2. there exists $C > 0$ such that

$$\begin{cases} \|G(t, u) - G(t', u)\|_{L^2(I)} \le C|t - t'|, \\ \forall t, t' \in [0, T], \forall u \in V \ s.t. \ \|u\|_V \le R. \end{cases} \tag{39}$$

These results come directly from (15).

Concerning the memory term, we have a similar result:

Lemma 2 *Take $T > 0$. Then*

$$\mathcal{F} : [0, T] \times C([-\tau, T]; L^2(I)) \to L^2(I), \quad \mathcal{F}(t, u)(x) := f\left(\int_{-\tau}^0 k(s, x) u(t+s)(x) ds \right)$$

is well defined. Moreover, we have the following estimates:

• *there exists $C > 0$ such that*

$$\begin{cases} \|\mathcal{F}(t, u)\|_{L^2(I)} \le C\|f\|_\infty, \\ \forall t \in [0, T], \forall u \in C([-\tau, T], L^2(I)), \end{cases} \tag{40}$$

and

$$\begin{cases} \|\mathcal{F}(t, u) - \mathcal{F}(t, v)\|_{L^2(I)} \le C\|u - v\|_{C([-\tau, T], L^2(I))}, \\ \forall t \in [0, T], \forall u, v \in C([-\tau, T], L^2(I)); \end{cases} \tag{41}$$

- *there exists $C > 0$ such that*

$$\begin{cases} \|\mathcal{F}(t, u) - \mathcal{F}(t', u)\|_{L^2(I)} \le C\|u^{(t)} - u^{(t')}\|_{C([-\tau, 0], L^2(I))}, \\ \forall t, t' \in [0, T], \forall u \in C([-\tau, T], L^2(I)). \end{cases} \tag{42}$$

6.2.2   Step 1: Γ Maps $\mathcal{X}_R$ into Itself if $t^* > 0$ Is Sufficiently Small

We recall that

$$\forall t \in [0, t^*], \quad \Gamma(u)(t) = e^{t\tilde{A}} u_0(0) + \int_0^t e^{(t-s)\tilde{A}} \left[\tilde{G}(s, u(s)) + F(u^{(s)}) \right] ds$$
$$= e^{t\tilde{A}} u_0(0) + \int_0^t e^{(t-s)\tilde{A}} \left[\tilde{G}(s, u(s)) + \mathcal{F}(s, u) \right] ds.$$

Denote

$$U_1(t) := e^{t\tilde{A}} u_0(0),$$
$$U_2(t) := \int_0^t e^{(t-s)\tilde{A}} \tilde{G}(s, u(s)) \, ds,$$
$$U_3(t) := \int_0^t e^{(t-s)\tilde{A}} \mathcal{F}(s, u) \, ds.$$

Then we claim that

$$U_1, U_2, U_3 \in C([0, t^*]; V), \quad U_1(0) = u_0(0), \quad U_2(0) = 0 = U_3(0). \tag{43}$$

Indeed: from standard regularity arguments, it is clear that $U_1(t), U_2(t), U_3(t) \in V$ for all $t \in (0, t^*]$. Moreover, $U_1(0) = u_0(0)$, $U_2(0) = 0 = U_3(0)$. It remains to show the continuity. If $t, t + h \in [0, t^*]$, then

- first

$$\begin{aligned} \|U_1(t+h) - U_1(t)\|_V &= \left\| (-\tilde{A})^{1/2} (e^{(t+h)\tilde{A}} - e^{t\tilde{A}}) u_0(0) \right\|_{L^2(I)} \\ &= \left\| (e^{(t+h)\tilde{A}} - e^{t\tilde{A}})((-\tilde{A})^{1/2} u_0(0)) \right\|_{L^2(I)} \\ &= \left\| \left(\int_t^{t+h} \tilde{A} e^{\sigma\tilde{A}} \, d\sigma \right) ((-\tilde{A})^{1/2} u_0(0)) \right\|_{L^2(I)} \\ &= \left\| \left(\int_t^{t+h} (-\tilde{A})^{3/4} e^{\sigma\tilde{A}} \, d\sigma \right) ((-\tilde{A})^{3/4} u_0(0)) \right\|_{L^2(I)} \\ &\le \left(\int_t^{t+h} \frac{c}{\sigma^{3/4}} \, d\sigma \right) \|(-\tilde{A})^{3/4} u_0(0)\|_{L^2(I)} \\ &\le c((t+h)^{1/4} - t^{1/4}) \|u_0(0)\|_{D(A)}, \end{aligned}$$

which gives that $U_1 \in C([0, t^*]; V)$;

- next, assume that $h > 0$, to simplify; then we have

$$
\begin{aligned}
\|U_2(t+h) &- U_2(t)\|_V \\
&= \left\|(-\tilde{A})^{1/2}\left(\int_0^{t+h} e^{(t+h-s)\tilde{A}}\tilde{G}(s, u(s))\, ds - \int_0^t e^{(t-s)\tilde{A}}\tilde{G}(s, u(s))\, ds\right)\right\|_{L^2(I)} \\
&= \left\|(-\tilde{A})^{1/2}(e^{h\tilde{A}} - Id)\int_0^t e^{(t-s)\tilde{A}}\tilde{G}(s, u(s))\, ds\right. \\
&\qquad\qquad \left. + (-\tilde{A})^{1/2}\int_t^{t+h} e^{(t+h-s)\tilde{A}}\tilde{G}(s, u(s))\, ds\right\|_{L^2(I)} \\
&\leq \left\|(e^{h\tilde{A}} - Id)\int_0^t (-\tilde{A})^{1/2}e^{(t-s)\tilde{A}}\tilde{G}(s, u(s))\, ds\right\|_{L^2(I)} \\
&\qquad\qquad + \left\|\int_t^{t+h}(-\tilde{A})^{1/2}e^{(t+h-s)\tilde{A}}\tilde{G}(s, u(s))\, ds\right\|_{L^2(I)};
\end{aligned}
$$

then, using once again (35), we have

$$
\begin{aligned}
\left\|(e^{h\tilde{A}} - Id)\int_0^t (-\tilde{A})^{1/2}e^{(t-s)\tilde{A}}\tilde{G}(s, u(s))\, ds\right\|_{L^2(I)} \\
= \left\|\left(\int_0^h \tilde{A}e^{\tau\tilde{A}}\, d\tau\right)\int_0^t (-\tilde{A})^{1/2}e^{(t-s)\tilde{A}}\tilde{G}(s, u(s))\, ds\right\|_{L^2(I)} \\
= \left\|\left(\int_0^h (-\tilde{A})^{3/4}e^{\tau\tilde{A}}\, d\tau\right)\left(\int_0^t (-\tilde{A})^{3/4}e^{(t-s)\tilde{A}}\tilde{G}(s, u(s))\, ds\right)\right\|_{L^2(I)} \\
\leq C\left(\int_0^h \frac{1}{\tau^{3/4}}\, d\tau\right)\left(\int_0^t \frac{1}{(t-s)^{3/4}}C_R(1 + R)\, ds\right) \\
= 16CC_R(1 + R)h^{1/4}t^{1/4};
\end{aligned}
$$

in the same way, using (34) we have

$$
\begin{aligned}
\left\|\int_t^{t+h}(-\tilde{A})^{1/2}\, e^{(t+h-s)\tilde{A}}\tilde{G}(s, u(s))\, ds\right\|_{L^2(I)} \\
\leq \int_t^{t+h} \frac{c}{\sqrt{t+h-s}}\|\tilde{G}(s, u(s))\|_{L^2(I)}\, ds \\
\leq C_R(1 + R)\int_t^{t+h} \frac{c}{\sqrt{t+h-s}}\, ds \\
= C_R(1 + R)O(\sqrt{|h|});
\end{aligned}
$$

hence

$$
\|U_2(t+h) - U_2(t)\|_V \leq C_R(1 + R)O(|h|^{1/4}),
$$

which gives that $u_2 \in C([0, t^*]; V)$;

- finally, using (34), (35) and (40) we have (still assuming that $h > 0$, in order to simplify)

$$\|U_3(t+h) - U_3(t)\|_V$$
$$= \left\|(-\tilde{A})^{1/2}\left(\int_0^{t+h} e^{(t+h-s)\tilde{A}}\mathcal{F}(s,u)\,ds - \int_0^t e^{(t-s)\tilde{A}}\mathcal{F}(s,u)\,ds\right)\right\|_{L^2(I)}$$
$$\leq \left\|(e^{h\tilde{A}} - Id)\int_0^t (-\tilde{A})^{1/2}e^{(t-s)\tilde{A}}\mathcal{F}(s,u)\,ds\right\|_{L^2(I)}$$
$$+ \left\|\int_t^{t+h}(-\tilde{A})^{1/2}e^{(t+h-s)\tilde{A}}\mathcal{F}(s,u)\,ds\right\|_{L^2(I)}$$
$$\leq \|f\|_\infty O(|h|^{1/4}),$$

which gives that $U_3 \in C([0, t^*]; V)$.

We conclude that $\Gamma(u) \in C([0, t^*]; V)$, and since $\Gamma(u)(0^+) = \Gamma(u)(0^-)$, we have that $\Gamma(u) \in C([-\tau, t^*]; V)$.

From the previous study, we also have that

$$\forall t \in [0, t^*], \quad \|\Gamma u(t)\|_V \leq \|U_1(t)\|_V + \|U_2(t)\|_V + \|U_3(t)\|_V$$
$$\leq \|u_0(0)\|_V + \int_0^t \frac{c}{\sqrt{t-s}}\left(C_R(1+R) + \|f\|_\infty\right)ds$$
$$= \|u_0(0)\|_V + \left(C_R(1+R) + \|f\|_\infty\right)O(\sqrt{t^*}).$$

Choose, e.g.,

$$R = \|u_0\|_{C[-\tau, 0]; V)} + 1;$$

then choosing t^* small enough, we have that

$$\forall t \in [-\tau, t^*], \quad \|\Gamma u(t)\|_V \leq R,$$

hence $\mathcal{X}_R$ is stable under the action of Γ if $t > 0$ is small enough.

6.2.3 Step 2: Γ Is a Contraction if $t^* \in (0, 1)$ Is Sufficiently Small

Of course,

$$\forall u, v \in \mathcal{X}_R, \forall t \in [-\tau, 0], \quad \Gamma(u)(t) = \Gamma(v)(t).$$

So we study the difference $\Gamma(u)(t) - \Gamma(v)(t)$ for $t \in [0, t^*]$. As we did previously, we have

$$\forall t \in [0, t^*], \quad \Gamma(u)(t) - \Gamma(v)(t) = \int_0^t e^{(t-s)\tilde{A}}\left[\tilde{G}(s, u(s)) - \tilde{G}(s, v(s))\right]ds$$
$$+ \int_0^t e^{(t-s)\tilde{A}}\left[\mathcal{F}(s, u) - \mathcal{F}(s, v)\right]ds.$$

Then, using (34) and (38), we have

$$
\begin{aligned}
&\left\| \int_0^t e^{(t-s)\tilde{A}} \left[\tilde{G}(s, u(s)) - \tilde{G}(s, v(s)) \right] ds \right\|_V \\
&= \left\| \int_0^t (-\tilde{A})^{1/2} e^{(t-s)\tilde{A}} \left[\tilde{G}(s, u(s)) - \tilde{G}(s, v(s)) \right] ds \right\|_{L^2(I)} \\
&\leq \int_0^t \frac{c}{\sqrt{t-s}} C_R \|u(s) - v(s)\|_V \, ds \\
&\leq C_R \|u - v\|_{C([-\tau, T]; V)} \int_0^t \frac{c}{\sqrt{t-s}} \, ds \\
&\leq 2c C_R \sqrt{t^*} \|u - v\|_{C([-\tau, T]; V)}.
\end{aligned}
$$

And using (34) and (41), we have

$$
\begin{aligned}
&\left\| \int_0^t e^{(t-s)\tilde{A}} \left[\mathcal{F}(s, u) - \mathcal{F}(s, v) \right] ds \right\|_V \\
&= \left\| \int_0^t (-\tilde{A})^{1/2} e^{(t-s)\tilde{A}} \left[\mathcal{F}(s, u) - \mathcal{F}(s, v) \right] ds \right\|_{L^2(I)} \\
&\leq \int_0^t \frac{c}{\sqrt{t-s}} C \|u - v\|_{C([-\tau, T]; V)} \, ds \\
&\leq 2c C_R \sqrt{t^*} \|u - v\|_{C([-\tau, T]; V)}.
\end{aligned}
$$

We obtain that

$$
\forall t \in [0, t^*], \quad \|\Gamma(u)(t) - \Gamma(v)(t)\|_V \leq 4c C_R \sqrt{t^*} \|u - v\|_{C([-\tau, T]; V)},
$$

hence Γ is a contraction if t^* is small enough.

6.2.4 Step 3: Additional Regularity of the Solution and Conclusion of the Proof of Proposition 2

Let us recall that $u \in C([-\tau, t^*]; V)$ and

$$
u(t) = e^{t\tilde{A}} u_0(0) + \int_0^t e^{(t-s)\tilde{A}} \tilde{G}(s, u(s)) ds + \int_0^t e^{(t-s)\tilde{A}} \mathcal{F}(s, u) \, ds.
$$

Let us remark that

$$
t \mapsto \int_0^t e^{(t-s)\tilde{A}} \tilde{G}(s, u(s)) ds \in H^1(0, t^*; L^2(I)) \cap L^2(0, t^*; D(A))
$$

and

$$
t \mapsto \int_0^t e^{(t-s)\tilde{A}} \mathcal{F}(s, u) \, ds \in H^1(0, t^*; L^2(I)) \cap L^2(0, t^*; D(A))
$$

using standard regularity results. Then we can conclude that

$$u \in H^1(0, t^*; L^2(I)) \cap L^2(0, t^*; D(A)) \cap C([-\tau, t^*]; V).$$

This concludes the proof of Proposition 2. □

6.3 Proof of Theorem 2: Uniqueness

In this section we prove the following

Proposition 3 *Given $T_0 > 0$ and $u_0 \in C([-\tau, 0]; V)$, assume that u and $\tilde{u}$ are mild solutions of the problem (20) on $[0, T_0]$. Then $u = \tilde{u}$ on $[0, T_0]$.*

Proof Consider $w = u - \tilde{u}$. Then w solves

$$\begin{cases} w_t - (\rho w_x)_x = G(t, u) - G(t, \tilde{u}) + f(H) - f(\tilde{H}), & t \in (0, T_0), x \in (-1, 1), \\ \rho w_x = 0, & t \in (0, T_0), x = \pm 1, \\ w(s, x) = 0, & s \in [-\tau, 0], x \in (-1, 1). \end{cases}$$

Take $T \in (0, T_0)$. Multiplying the first equation by w and integrating on $(0, T) \times (-1, 1)$, we obtain

$$\int_0^T \int_{-1}^1 w(w_t - (\rho w_x)_x) = \int_0^T \int_{-1}^1 w(G(t, u) - G(t, \tilde{u}) + f(H) - f(\tilde{H})).$$

Integrating by parts, we have

$$\int_0^T \int_{-1}^1 w w_t \, dx \, dt = \frac{1}{2} \|w(T)\|_{L^2(-1,1)}^2,$$

$$\int_0^T \int_{-1}^1 w(-(\rho w_x)_x)) \, dx \, dt \geq 0,$$

$$\int_0^T \int_{-1}^1 w(G(t, u) - G(t, \tilde{u})) \, dx \, dt \leq c \int_0^T \int_{-1}^1 w(u - \tilde{u}) \, dx \, dt$$

$$= c \int_0^T \|w(t)\|_{L^2(-1,1)}^2 \, dt,$$

and

$$\int_0^T \int_{-1}^1 w(f(H) - f(\tilde{H})) \, dx \, dt \le c \int_0^T \int_{-1}^1 |w||H - \tilde{H}| \, dx \, dt$$

$$\le c \int_0^T \|w(t)\|_{L^2(-1,1)}^2 \, dt + c \int_0^T \int_{-1}^1 |H - \tilde{H}|^2 \, dx \, dt.$$

Let us introduce

$$W(T') := \int_0^{T'} \|w(T)\|_{L^2(-1,1)}^2 \, dT.$$

Using the previous estimates, we have

$$\frac{1}{2} W'(T) \le 2cW(T) + c \int_0^T \int_{-1}^1 |H - \tilde{H}|^2 \, dx \, dt. \tag{44}$$

Concerning the last term:

$$\int_0^T \int_{-1}^1 |H - \tilde{H}|^2 \, dx \, dt$$

$$= \int_0^T \int_{-1}^1 \left| \int_{-\tau}^0 k(s, x)(u(t + s, x) - \tilde{u}(t + s, x)) \, ds \right|^2 dx \, dt$$

$$\le c \int_0^T \int_{-1}^1 \left(\int_{-\tau}^0 (u(t + s, x) - \tilde{u}(t + s, x))^2 \, ds \right) dx \, dt;$$

but

$$\int_0^T \int_{-1}^1 \left(\int_{-\tau}^0 (u(t + s, x) - \tilde{u}(t + s, x))^2 \, ds \right) dx \, dt$$

$$= \int_0^T \int_{-\tau}^0 \|w(t + s)\|_{L^2(-1,1)}^2 \, ds \, dt$$

$$= \int_0^T \int_{t-\tau}^t \|w(\sigma)\|_{L^2(-1,1)}^2 \, d\sigma \, dt.$$

Note that the initial condition of w gives us that: first, if $t - \tau \le 0$ then

$$\int_{t-\tau}^t \|w(\sigma)\|_{L^2(-1,1)}^2 \, d\sigma = \int_0^t \|w(\sigma)\|_{L^2(-1,1)}^2 \, d\sigma,$$

and of course, if $t - \tau \ge 0$ then

$$\int_{t-\tau}^t \|w(\sigma)\|_{L^2(-1,1)}^2 \, d\sigma \le \int_0^t \|w(\sigma)\|_{L^2(-1,1)}^2 \, d\sigma.$$

Hence

$$\int_0^T \int_{t-\tau}^t \|w(\sigma)\|_{L^2(-1,1)}^2 \, d\sigma \, dt \le \int_0^T \int_0^t \|w(\sigma)\|_{L^2(-1,1)}^2 \, d\sigma \, dt,$$

which gives that

$$\int_0^T \int_{-1}^1 |H - \tilde{H}|^2 \, dx \, dt \le c \int_0^T \int_0^t \|w(\sigma)\|_{L^2(-1,1)}^2 \, d\sigma \, dt = c \int_0^T W(t) \, dt.$$

Since W is nondecreasing, we obtain that

$$\int_0^T \int_{-1}^1 |H - \tilde{H}|^2 \, dx \, dt \le cT W(T), \tag{45}$$

and then we are in position to conclude: we deduce from (44) and (45) that

$$\frac{1}{2} W'(T) \le (2c + c'T) W(T).$$

Finally, integrating with respect to $T \in (0, T')$, and using that $W(0) = 0$, we obtain that

$$W(T') \le 2 \int_0^{T'} (2c + c'T) W(T) \, dT.$$

Then Gronwall's lemma tells us that $W = 0$, and then $W' = 0$, which gives that $w = 0$. This concludes the proof of Proposition 3.

6.4 Proof of Theorem 2: The Maximal Solution Is Global in Time

Proposition 3 is a standard uniqueness result. As a consequence, combining it with the local existence result given in Proposition 2, we are able to define the maximal existence time:

$$T^*(u_0) := \sup\{T \ge 0 \text{ s.t. (20) has a mild solution on } [0, T]\}, \tag{46}$$

Then, for all $T < T^*(u_0)$, we have a mild solution u_T on $[0, T]$, and if $0 < T < T' < T^*(u_0)$, then

$$u_T = u_{T'} \text{ on } [0, T],$$

and this allows us to define the associated maximal solution, defined exactly on $[0, T^*(u_0))$. It remains to prove the global existence of the maximal solution:

Proposition 4 *Consider $u_0 \in C([-\tau, 0]; V)$ and such that $u_0(0) \in D(A) \cap L^\infty(I)$, and the associated maximal mild solution, defined in $[0, T^*(u_0))$. Then $T^*(u_0) = +\infty$.*

Proposition 4 allows us to conclude the proof of Theorem 2. So it remains to prove Proposition 4, and for this we begin by proving a boundedness property.

6.4.1 L^∞ Bound of the Solution

First of all, let us prove a general and useful boundedness property. Following the proofs developed by Tort and Vancostenoble [42], without the memory term, we first state a known preliminary result (see Lemma 6.1 in [42]):

Lemma 3 *Let $u \in V$. Then, for all M, $(u - M)^+ := \sup(u - M, 0) \in V$ and $(u + M)^- := \sup(-(u + M), 0) \in V$. Moreover for a.e. $x \in I$*

$$((u - M)^+)_x(x) = \begin{cases} u_x(x) & (u - M)(x) > 0 \\ 0 & (u - M)(x) \le 0 \end{cases} \tag{47}$$

and for a.e. $x \in I$

$$((u + M)^-)_x(x) = \begin{cases} 0 & (u + M)(x) > 0 \\ -u_x(x) & (u + M)(x) \le 0 \end{cases}. \tag{48}$$

Then we can prove the theorem below, where we just add the memory term to the proof of Theorem 3.3 in [42]:

Theorem 6 *Consider $u_0 \in C([-\tau, 0]; V)$ and $u_0(0) \in D(A) \cap L^\infty(I)$, $T > 0$ and u a mild solution of (20) defined on $[0, T]$. Let us denote*

$$M_1 := \left(\frac{||q||_{L^\infty(I)} ||r||_{L^\infty(-\infty, +\infty)} ||\beta||_{L^\infty(-\infty, +\infty)} + ||f||_{L^\infty(-\infty, +\infty)}}{\varepsilon_1} \right)^{\frac{1}{4}}$$

and

$$M := \max\{||u_0(0)||_{L^\infty(I)}, M_1\}.$$

Then u satisfies

$$||u||_{L^\infty((0, T) \times I)} \le M. \tag{49}$$

Proof Let us set $\mathcal{B} := \{x \in I : u(t, x) > M\}$ and multiply the equation satisfied by u by $(u - M)^+$, then we get the equation below using the previous Lemma and the boundary conditions satisfied by u.

$$\int_I u_t(u - M)^+ dx + \int_I \rho(((u - M)^+)^2)_x dx$$
$$= \int_I [Q\beta(u) - R_e(u) + F(u^{(t)})](u - M)^+ dx$$
$$= \int_{\mathcal{B}} [Q\beta(u) - R_e(u) + F(u^{(t)})](u - M)dx.$$

Moreover, for $x \in \mathcal{B}$,

$$Q\beta(u) - \varepsilon(u)u|u|^3 + F(u^{(t)}) \leq \|Q\|_\infty\|\beta\|_\infty - \varepsilon_1 M^4 + \|f\|_\infty \leq 0 \quad (50)$$

thanks to our definition of M and to the assumptions on F. Then

$$\frac{1}{2}\frac{d}{dt}\int_I |(u - M)^+|^2 dx = \int_I u_t(u - M)^+ dx \leq 0.$$

for all $t \in [0, T]$. Therefore $t \mapsto \|(u - M)^+(t)\|^2_{L^2(I)}$ is nonincreasing on $[0, T]$. Since $(u_0(0) - M)^+ \equiv 0$, we obtain that $u(t, x) \leq M$ for all $t \in [0, T]$ and for a.e. $x \in I$.

In the same way, we can multiply Eq. (1) by $(u + M)^-$ and then we obtain

$$\frac{d}{dt}\int_I |(u + M)^-|^2 dx \leq 0.$$

Finally, since $(u_0(0) + M)^- \equiv 0$ we have that $u(t, x) \geq -M$ for all $t \in [0, T]$ and a.e. $x \in I$.

6.4.2 Proof of Proposition 4

From the previous theorem one may deduce that, for any $u_0(0) \in D(A) \cap L^\infty(I)$, the L^∞-norm of the solution remains bounded on $[0, T^*(u_0))$. To ensure the global existence of the mild solution for the Sellers-type model, we argue by contradiction: we are going to prove that, if $T^*(u_0) < +\infty$, then $t \mapsto u(t)$ can be extended up to $T^*(u_0)$ (and then further), which will be in contradiction with the maximality of $T^*(u_0)$.

Let us assume that $T^*(u_0) < +\infty$. Note that since $u_0(0) \in D(A) \cap L^\infty(I)$, Theorem 6 implies that $\|u\|_{L^\infty((0,T^*(u_0))\times I)} \leq M$. It follows that

$$R_e(u) = \varepsilon(u)u|u|^3 \leq \|\varepsilon\|_{L^\infty(-\infty,+\infty)} M^4 =: C.$$

It remains to prove that there exists $\lim_{t \uparrow T^*(u_0)} u(t)$ in V. To prove this, we prove that the function $t \mapsto u(t)$ satisfies the Cauchy criterion. By definition of mild solution, we have

$$u(t) = U_1(t) + U_2(t) + U_3(t)$$

with

$$
\begin{aligned}
U_1(t) &:= e^{t\tilde{A}} u_0(0), \\
U_2(t) &:= \int_0^t e^{(t-s)\tilde{A}} \tilde{G}(s, u(s)) \, ds, \\
U_3(t) &:= \int_0^t e^{(t-s)A} \mathcal{F}(s, u) \, ds.
\end{aligned}
$$

Now, first, U_1 has a limit in V: $U_1(t) \to U_1(t^*)$ as $t \to t^*$ since it is a semigroup applied to the initial value.

Next, let us prove that U_2 has also a limit in V as $t \to t^*$: if $t' \leq t < t^*$, we have

$$
\begin{aligned}
\|U_2(t) - U_2(t')\|_V &= \left\| \int_0^t e^{(t-s)\tilde{A}} \tilde{G}(s, u(s)) \, ds - \int_0^{t'} e^{(t'-s)\tilde{A}} \tilde{G}(s, u(s)) \, ds \right\|_V \\
&\leq \left\| \int_0^{t'} e^{(t-s)\tilde{A}} \tilde{G}(s, u(s)) - e^{(t'-s)\tilde{A}} \tilde{G}(s, u(s)) \, ds \right\|_V + \left\| \int_{t'}^t e^{(t-s)\tilde{A}} \tilde{G}(s, u(s)) \, ds \right\|_V.
\end{aligned}
$$

We study these two last terms, proving that they satisfy the Cauchy criterion, hence both will have a limit in V:

- first the last one: using Theorem 6, we have for all $s \in [0, T^*(u_0))$:

$$
\begin{aligned}
\|\tilde{G}(s, u(s))\|_{L^2(I)} &= \|r(s)q(x)\beta(u) - \varepsilon(u)u|u|^3\|_{L^2(I)} \\
&\leq 2\Big(\|r\|_\infty \|q\|_\infty \|\beta\|_\infty + \|\varepsilon\|_\infty M^4 \Big),
\end{aligned}
$$

hence

$$
\begin{aligned}
\left\| \int_{t'}^t e^{(t-s)\tilde{A}} \tilde{G}(s, u(s)) \, ds \right\|_V &= \left\| (-\tilde{A})^{1/2} \int_{t'}^t e^{(t-s)\tilde{A}} \tilde{G}(s, u(s)) \, ds \right\|_{L^2(I)} \\
&\leq \int_{t'}^t \left\| (-\tilde{A})^{1/2} e^{(t-s)\tilde{A}} \tilde{G}(s, u(s)) \right\|_{L^2(I)} ds \leq \int_{t'}^t \frac{C}{\sqrt{t-s}} \, ds = 2C\sqrt{t - t'};
\end{aligned}
$$

- for the other term:

$$\left\| \int_0^{t'} e^{(t-s)A} G(s, u(s)) - e^{(t'-s)A} G(s, u(s)) \, ds \right\|_V$$

$$= \left\| (e^{(t-t')A} - Id) \int_0^{t'} e^{(t'-s)A} G(s, u(s)) \, ds \right\|_V$$

$$= \left\| (-A)^{1/2} (e^{(t-t')A} - Id) \int_0^{t'} e^{(t'-s)A} G(s, u(s)) \, ds \right\|_{L^2(I)}$$

$$= \left\| (e^{(t-t')A} - Id) \int_0^{t'} (-A)^{1/2} e^{(t'-s)A} G(s, u(s)) \, ds \right\|_{L^2(I)}$$

$$= \left\| \left(\int_0^{t-t'} A e^{\tau A} \, d\tau \right) \left(\int_0^{t'} (-A)^{1/2} e^{(t'-s)A} G(s, u(s)) \, ds \right) \right\|_{L^2(I)}$$

$$= \left\| \left(\int_0^{t-t'} (-A)^{3/4} e^{\tau A} \, d\tau \right) \left(\int_0^{t'} (-A)^{3/4} e^{(t'-s)A} G(s, u(s)) \, ds \right) \right\|_{L^2(I)};$$

using (35) and the fact that G is bounded, we obtain that

$$\left\| \left(\int_0^{t-t'} (-A)^{3/4} e^{\tau A} \, d\tau \right) \left(\int_0^{t'} (-A)^{3/4} e^{(t'-s)A} G(s, u(s)) \, ds \right) \right\|_{L^2(I)}$$

$$\leq C \left(\int_0^{t-t'} \frac{1}{\tau^{3/4}} \, d\tau \right) \left(\int_0^{t'} \frac{1}{(t'-s)^{3/4}} \, ds \right) = C'(t - t')^{1/4} (t')^{1/4}.$$

From these two estimates, we deduce that $t \mapsto U_2(t)$ satisfies the Cauchy criterion and has a limit as $t \to T^*(u_0)$ if $T^*(u_0) < +\infty$.

In the same way, $t \mapsto U_3(t)$ satisfies the Cauchy criterion as $t \to T^*(u_0)$. It follows that $t \mapsto u(t)$ has a limit as $t \to T^*(u_0)$ if $T^*(u_0) < +\infty$, which contradicts the maximality of $T^*(u_0)$, hence $T^*(u_0) = +\infty$. $\quad\square$

Remark 1 If we assume that R_e is linear as in the Budyko models, i.e. $R_e(u) = A + Bu$, everything remains true, and we can apply the fixed point theorem and extend the solution to a global one.

7 Proof of Theorem 3

Without loss of generality, we can assume that

$$0 < T < \delta.$$

Using the extra assumption (24), we see that the history term satisfies

$$\forall t \in (0, T), \quad H(t, x, u) = \int_{-\tau}^{0} k(s, x)u(t + s, x)ds$$

$$= \int_{-\tau}^{-\delta} k(s, x)u(t + s, x)ds = \int_{t-\tau}^{t-\delta} k(\sigma - t, x)u(\sigma, x)d\sigma$$

$$= \int_{t-\tau}^{t-\delta} k(\sigma - t, x)u_0(\sigma, x)d\sigma = H(t, x, u_0),$$

where we used (10) and that $[t - \tau, t - \delta] \subset [-\tau, 0]$ since $t \leq T < \delta$.

Hence during this small interval of time, the memory term depends only on the initial condition, hence

$$H(t, x, u) = H(t, x, u_0) = H(t, x, \tilde{u}). \tag{51}$$

Let us set $v := u - \tilde{u}$.

7.1 Step 1: The Linear Problem Satisfied by v

Subtracting the problem (1) satisfied by u with the one satisfied by $\tilde{u}$, we obtain that the function v verifies

$$v_t - (\rho(x)v_x)_x = r(t)q(x)\beta(u) - r(t)\tilde{q}(x)\beta(\tilde{u}) - (\varepsilon(u)|u|^3 u - \varepsilon(\tilde{u})|\tilde{u}|^3 \tilde{u}) \tag{52}$$

for all $t \in (0, T)$ and $x \in (-1, 1)$. We linearize (52) thanks to the regularity of β (in Sellers type models) and of ε, defining

$$\mu_1(u, \tilde{u}) \quad := \begin{cases} \frac{\varepsilon(u)|u|^3 u - \varepsilon(\tilde{u})|\tilde{u}|^3 \tilde{u}}{u - \tilde{u}} & u \neq \tilde{u} \\ \frac{\partial}{\partial u}(\varepsilon(u)|u|^3 u) & u = \tilde{u} \end{cases}$$

$$\text{and} \quad \mu_2(u, \tilde{u}) := \begin{cases} \frac{\beta(u) - \beta(\tilde{u})}{u - \tilde{u}} & u \neq \tilde{u} \\ \frac{\partial}{\partial u}(\beta(u)) & u = \tilde{u} \end{cases}$$

Let us add and subtract $r(t)\tilde{q}(x)\beta(u)$ and then replace μ_1 and μ_2 in (52), so we obtain the following linear equation with respect to v:

$$v_t - (\rho(x)v_x)_x = r(t)\tilde{q}(x)\mu_2(u, \tilde{u})v - \mu_1(u, \tilde{u})v + r(t)\beta(u)(q(x) - \tilde{q}(x)). \tag{53}$$

7.2   Step 2: $q = \tilde{q}$ on $(-1, x_0)$

We define the largest interval $[y_1, x_0]$ where $q \equiv \tilde{q}$ and we want to prove that $y_1 = -1$.

Let us set

$$\mathcal{A}^- := \{x \leq x_0 : (q - \tilde{q})(y) \equiv 0 \quad \forall y \in [x, x_0]\}$$

If $\mathcal{A}^- \neq \emptyset$, we consider

$$y_1 := \inf \mathcal{A}^-,$$

and if $\mathcal{A}^- = \emptyset$, we consider

$$y_1 := x_0,$$

so that in any case, we know that if $y_1 > -1$, and if $\eta > 0$ is such that $y_1 - \eta > -1$, then there exists $y_2 \in (y_1 - \eta, y_1)$ such that $q(y_2) \neq \tilde{q}(y_2)$.

To show that $y_1 = -1$, we argue by contradiction, so let us assume that $y_1 > -1$.

Step 2.1 First of all we want to prove that there exists $t_1 \in [0, T)$ and $y_2 \in (-1, y_1)$ such that $v(t, y_2)$ never vanishes on $(0, t_1)$. Since $q, \tilde{q} \in \mathcal{M}$, we have that $q - \tilde{q} \in \mathcal{M}$. It follows that there exists $y_1' < y_1$ such that $q - \tilde{q}$ is analytic on $[y_1', y_1]$, hence is constantly equal to zero or has only a finite number of zeros. Since we already noted that the definition of y_1 implies that $q - \tilde{q}$ cannot be constantly equal to zero on some interval $(y_1 - \eta, y_1]$ (with $\eta > 0$), then $q - \tilde{q}$ has only a finite number of zeros in $[y_1', y_1]$, and this implies that

$$\exists \, y_2 \in (y_1', y_1) \text{ s.t } (q - \tilde{q})(x) \neq 0 \text{ for all } x \in [y_2, y_1). \tag{54}$$

Without loss of generality, we can assume that

$$(q - \tilde{q})(x) > 0 \quad \forall x \in [y_2, y_1). \tag{55}$$

Let us notice that since u and $\tilde{u}$ have the same initial condition, we have that $v(0, x) = 0$ for all $x \in [-1, 1]$ and that $v_x(0, x) = 0$. Using this remark and computing (53) at $t = 0$ and $x = y_2$, we obtain

$$v_t(0, y_2) = r(0)\beta(u_0)(q - \tilde{q})(y_2),$$

and then (55) implies that $v_t(0, y_2) > 0$. Therefore, since $v(0, y_2) = 0$, we have that there exists some time $t_1 \in (0, T)$ such that

$$v(t, y_2) > 0 \quad \forall t \in (0, t_1).$$

Moreover, the assumption (28) implies that $v(t, x_0) = 0$ for all $t \in (0, T)$.

Step 2.2 Using the strong maximum principle and the Hopf's Lemma, we are going to prove that the assumption $y_1 > -1$ leads to a contradiction. Consider

$$K := \max_{t \in [0, t_1], x \in [y_2, x_0]} -\mu_1(u(t, x), \tilde{u}(t, x)) + r(t)\tilde{q}(x)\mu_2(u(t, x), \tilde{u}(t, x)) :$$

K is chosen so that

$$R(t, x) := -\mu_1(u(t, x), \tilde{u}(t, x)) + r(t)\tilde{q}(x)\mu_2(u(t, x), \tilde{u}(t, x)) - K \le 0.$$

Let also define

$$w(t, x) := v(t, x)e^{-Kt}.$$

Using (53), we observe that

$$w_t - (\rho(x)w_x)_x - R(t, x)w = r(t)\beta(u)(q(x) - \tilde{q}(x)) e^{-Kt}.$$

Since

$$\forall x \in [y_2, x_0], \quad q(x) - \tilde{q}(x) \ge 0,$$

we obtain that w satisfies

$$\begin{cases} w_t - (\rho(x)w_x)_x - R(t, x)w \ge 0, & (t, x) \in (0, t_1) \times (y_2, x_0), \\ w(0, x) = 0, & x \in [y_2, x_0], \\ w(t, x_0) - 0, & t \in (0, t_1), \\ w(t, y_2) > 0, & t \in (0, t_1) \end{cases} \tag{56}$$

where the second condition follows from the initial conditions of (1), the third from the assumption (28), and the last from Step 2.1.

Let us notice that, since $[y_2, x_0] \subset (-1, 1)$, we can apply the strong maximum principle (Chapter 3 of [36]). It implies that

$$w(t, x) > 0 \qquad \forall(t, x) \in (0, t_1) \times (y_2, x_0). \tag{57}$$

Moreover, since $w(t, x_0) = 0$ and $x_0 \ne 1$, we can apply the Hopf's Lemma which implies that

$$w_x(t, x_0) < 0 \qquad \forall t \in (0, t_1).$$

It follows that $u_x(t, x_0) < \tilde{u}_x(t, x_0)$ for all $t \in (0, t_1)$ which contradicts the second assumption in (28).

As a consequence, the assumption $y_1 > -1$ is false and this implies that $y_1 = -1$, therefore $q \equiv \tilde{q}$ on $(-1, x_0]$.

7.3 Conclusion

The proof is equivalent for $[x_0, 1)$, and one can show that $q \equiv \tilde{q}$ on $(-1, 1)$.
The uniqueness result of Theorem 2 implies that $u \equiv \tilde{u}$ on $[0, +\infty) \times (-1, 1)$. $\square$

8 Proof of Theorem 4

We follow the strategy used in [32, 42] (that was adapted from the method introduced in Imanuvilov and Yamamoto, to study the Sellers case), focusing on the changes brought by the memory term. The proof of the stability result is decomposed in several steps, we give the main intermediate results and we will refer to [32, 42] for some details.

Remember that u and $\tilde{u}$ are the solutions of (14) and (25). Thanks to the assumptions, we can assume that $T < \delta$ without loss of generality.

8.1 Step 1: The Problem Solved by the Difference $w := u - \tilde{u}$

Clearly the difference

$$w := u - \tilde{u} \tag{58}$$

satisfies the problem

$$\begin{cases} w_t - (\rho w_x)_x = K^* + K + \tilde{K} + K^h, & t > 0, x \in (-1, 1), \\ \rho w_x = 0, & x = \pm 1, \\ w(s, x) = u_0(s, x) - \tilde{u}_0(s, x), & s \in [-\tau, 0], x \in (-1, 1), \end{cases} \tag{59}$$

where the source terms K^*, K, $\tilde{K}$ and K^h are defined by

$$K^* := r(t)(q(x) - \tilde{q}(x))\beta(u), \tag{60}$$

$$K := r(t)\tilde{q}(x)(\beta(u) - \beta(\tilde{u})), \tag{61}$$

$$\tilde{K} := -\varepsilon(u)|u|^3 u + \varepsilon(\tilde{u})|\tilde{u}|^3 \tilde{u}, \tag{62}$$

$$K^h := f(H) - f(\tilde{H}). \tag{63}$$

Note that compared to [32, 42], our goal is similarly to estimate from above K^* and the only difference lies in the presence of K^h. However, since we assumed that the memory kernel k is supported in $[-\tau, -\delta]$, and that $T < \delta$, it is clear (as in the previous section) that the memory terms H and $\tilde{H}$ are directly determined from the initial conditions u_0 and $\tilde{u}_0$.

8.2   Step 2: A Useful Property of the Source Term K^*

We claim that K^* satisfies the following property (classical in that question of determining a source term):

$$\exists C_0 > 0 \quad \text{s.t.} \quad \forall t \in (0, T), \forall x \in (-1, 1), \quad \left|\frac{\partial K^*}{\partial t}(t, x)\right| \leq C_0 |K^*(T', x)|. \tag{64}$$

Since

$$K_t^* := r'(t)(q(x) - \tilde{q}(x))\beta(u) + r(t)(q(x) - \tilde{q}(x))\beta'(u)u_t,$$

(64) is an easy consequence of the following regularity result:

Lemma 4 *Under the regularity assumptions of Theorem 4, the solution u of (14) satisfies: $u_t \in L^\infty((0, T) \times I)$, and more precisely, there exists $C(T, M, M') > 0$ such that, for all $u_0 \in \mathcal{U}_M^{(loc)}$, for all $q \in \mathcal{Q}_{M'}^{(loc)}$, we have*

$$\|u_t\|_{L^\infty((0,T) \times I)} \leq C(T, M, M').$$

Lemma 4 can be proved as Theorem 3.4 and Corollary 3.1 of [42], noting that the additive memory term satisfies

$$\forall t \in (0, T), \quad H(t, x) = \int_{t-\tau}^{t-\delta} k(s-t, x)u(s, x)\, ds = \int_{t-\tau}^{t-\delta} k(s-t, x)u_0(s, x)\, ds,$$

hence

$$H_t = k(-\tau, x)u_0(t - \tau, x) - k(-\delta, x)u_0(t - \delta, x) - \int_{t-\tau}^{t-\delta} k_t(s - t, x)u_0(s, x)\, ds,$$

which is bounded from the regularity assumptions of Theorem 4.

8.3 Step 3: A Carleman Estimate on the Problem Solved by $z := w_t$

Consider $z := w_t$. It is solution of the following problem

$$\begin{cases} z_t - (\rho z_x)_x = K_t^* + K_t + \tilde{K}_t + K_t^h, & t > 0, x \in (-1, 1), \\ \rho z_x = 0, & x = \pm 1, \end{cases} \tag{65}$$

and we can apply standard Carleman estimates for such degenerate operator (see [7, 9]): choosing

- $\theta : (t_0, T) \to (0, +\infty)$ smooth, strictly convex, such that

$$\theta(t) \to +\infty \quad \text{as } t \to t_0^+ \text{ and as } t \to T^-,$$

and $\theta'(T') = 0$ such that T' is the point of global minimum,
- and $p : (-1, 1) \to [1, +\infty)$ well designed with the respect to the degeneracy (see [7] in the typical degenerate case, and [42] for an explicit construction in the case of the Sellers model),

and considering

$$\sigma(t, x) = \theta(t) p(x),$$

and $R > 0$ large enough, the following Carleman estimate holds true, see Theorem 4.2 in [42]:

$$\int_{t_0}^T \int_{-1}^1 \left(R^3 \theta^3 (1 - x^2) z^2 + R\theta (1 - x^2) z_x^2 + \frac{1}{R\theta} z_t^2 \right) e^{-2R\sigma}$$
$$\leq C_1 \left(\int_{t_0}^T \int_{-1}^1 (K_t^* + K_t + \tilde{K}_t + K_t^h)^2 e^{-2R\sigma} + \int_{t_0}^T \int_a^b R^3 \theta^3 z^2 e^{-2R\sigma} \right). \tag{66}$$

Then,

- since u_t is bounded, we immediately obtain that

$$|K_t| + |\tilde{K}_t| \leq c(|w| + |z|),$$

and this allows to show that

$$\int_{t_0}^T \int_{-1}^1 (K_t^2 + \tilde{K}_t^2) e^{-2R\sigma} \leq c \left(\|w(T')\|_{L^2(I)}^2 + \int_{t_0}^T \int_{-1}^1 z^2 e^{-2R\sigma} \right); \tag{67}$$

- for the memory term: using the explicit form given by the initial conditions, we immediately have

$$K_t^h = f'(H)H_t - f'(\tilde{H})\tilde{H}_t = f'(H)(H_t - \tilde{H}_t) + (f'(H) - f'(\tilde{H}))\tilde{H}_t,$$

hence

$$|K_t^h| \le c\Big(|H_t - \tilde{H}_t| + |H - \tilde{H}|\Big)$$

and then

$$\int_{t_0}^T \int_{-1}^1 K_h^2 e^{-2R\sigma} \le c\|u_0 - \tilde{u}_0\|_{C([-\tau,0],V)}^2. \tag{68}$$

Using (67) and (68) in the Carleman estimate (66), we obtain

$$\int_{t_0}^T \int_{-1}^1 \Big(R^3\theta^3(1-x^2)z^2 + R\theta(1-x^2)z_x^2 + \frac{1}{R\theta}z_t^2\Big)e^{-2R\sigma}$$
$$\le C\Big(\int_{t_0}^T \int_{-1}^1 (K_t^*)^2 e^{-2R\sigma} + \int_{t_0}^T \int_a^b R^3\theta^3 z^2 e^{-2R\sigma}\Big)$$
$$+C\Big(\|w(T')\|_{L^2(I)}^2 + \int_{t_0}^T \int_{-1}^1 z^2 e^{-2R\sigma}\Big) + c\|u_0 - \tilde{u}_0\|_{C([-\tau,0],V)}^2.$$

Absorbing the term $\int_{t_0}^T \int_{-1}^1 z^2 e^{-2R\sigma}$ in the left-hand side (this is classical, using Hardy type inequalities, see [7, 42]), we finally obtain

$$I_0 := \int_{t_0}^T \int_{-1}^1 \Big(R^3\theta^3(1-x^2)z^2 + R\theta(1-x^2)z_x^2 + \frac{1}{R\theta}z_t^2\Big)e^{-2R\sigma}$$
$$\le C\int_{t_0}^T \int_{-1}^1 (K_t^*)^2 e^{-2R\sigma} + C\int_{t_0}^T \int_a^b R^3\theta^3 z^2 e^{-2R\sigma}\Big) \tag{69}$$
$$+C\|w(T')\|_{L^2(I)}^2 + C\|u_0 - \tilde{u}_0\|_{C([-\tau,0],V)}^2.$$

8.4 Step 4: An Estimate from Above

Using Step 2, we see that

$$\int_{t_0}^T \int_{-1}^1 (K_t^*)^2 e^{-2R\sigma}\, dx\, dt \le C_0^2 \int_{t_0}^T \int_{-1}^1 K^*(T')^2 e^{-2R\sigma}\, dx\, dt$$
$$= C_0^2 \int_{-1}^1 K^*(T')^2 \Big(\int_{t_0}^T e^{-2R\sigma}\, dt\Big)\, dx,$$

and it is classical (see Imanuvilov and Yamamoto [29, equation (3.17)]) that

$$\int_{t_0}^{T} e^{-2R\sigma} \, dt = o(e^{-2R\sigma(T')}) \quad \text{as } R \to +\infty.$$

(This is due to the convexity of the function θ, that attains its minimum at T'.) Hence

$$\int_{t_0}^{T} \int_{-1}^{1} (K_t^*)^2 e^{-2R\sigma} \, dx \, dt = o\Big(\int_{-1}^{1} K^*(T')^2 e^{-2R\sigma(T')} \, dx \Big) \quad \text{as } R \to \infty.$$

(70)

8.5   Step 5: An Estimate from Below

As in [32, 42], we have

$$\int_{-1}^{1} z(T')^2 e^{-2R\sigma(T')} \le cI_0.$$

(71)

8.6   Step 6: Conclusion

Now we are in position to conclude: using the equation in w:

$$K^*(T') = z(T') - (\rho w_x)_x(T') - K(T') - \tilde{K}(T') - K^h(T'),$$

hence

$$\int_{-1}^{1} K^*(T')^2 e^{-2R\sigma(T')} \le C\Big(\int_{-1}^{1} (z(T')^2 + (\rho w_x)_x(T')^2 \\ + K(T')^2 + \tilde{K}(T')^2 + K^h(T')^2) e^{-2R\sigma(T')} \Big);$$

now note that

$$|K(T')| \le c|w(T')|, \quad |\tilde{K}(T')| \le c|w(T')|,$$

and

$$\int_{-1}^{1} K^h(T')^2 e^{-2R\sigma(T')} \le C\|u_0 - \tilde{u}_0\|_{C([-\tau,0],V)}^2;$$

then using (71), (69) and (70), we obtain that

$$\int_{-1}^{1} K^*(T')^2 \, e^{-2R\sigma(T')}$$

$$\leq C\Big(\|w(T')\|_{D(A)}^2 + \|w_t\|_{L^2((t_0,T)\times(a,b))}^2 + \|u_0 - \tilde{u}_0\|_{C([-\tau,0];V)}^2\Big).$$

Looking to the form of K^*, we obtain (31). □

9 Proof of Theorem 5

9.1 The Strategy to Prove Theorem 5

The method is usual:

- first we approximate the coalbedo β by a sequence of smooth functions β_j,
- then we consider the approximate problem associated to β_j, and we denote u_j its (unique) solution,
- we obtain suitable assumptions on u_j, and we pass to the limit, and we prove that we obtain a solution u_∞ of the original problem.

Let us be more precise: first, given $j \geq 1$, we consider a function
$\beta_j : (-\infty, +\infty) \to (-\infty, +\infty)$ which is of class C^2, nondecreasing, and

$$\begin{cases} \beta_j(u) = a_i, & u \leq \bar{u} - \frac{1}{j}, \\ \beta_j(u) = a_f, & u \geq \bar{u} + \frac{1}{j}. \end{cases}$$

Then we can consider the approximate problem

$$\begin{cases} u_t - (\rho(x)u_x)_x = r(t)q(x)\beta_j(u) - (a + bu) + f(H), & t > 0, x \in I, \\ \rho(x)u_x = 0, & x = \pm 1, \\ u(s, x) = u_0(s, x), & s \in [-\tau, 0], x \in I, \end{cases}$$
$$(72)$$

which is of Sellers type. Hence, since $u_0 \in C([-\tau, 0]; V)$, $u_0(0) \in D(A) \cap L^\infty(I)$, Theorem 2 ensures us that, given $T > 0$, the problem (72) has a unique solution u_j such that, for all $T > 0$, we have

$$u_j \in H^1(0, T; L^2(I)) \cap L^2(0, T; D(A)) \cap C([-\tau, T]; V). \tag{73}$$

We will denote

$$\gamma_j(t, x) := r(t)q(x)\beta_j(u_j(t, x)) - (a + bu_j(t, x)) + f(H_j(t, x)), \tag{74}$$

where of course

$$H_j(t, x) = \int_{-\tau}^0 k(s, x), u_j(t + s, x)\, ds.$$

In order to pass to the limit $j \to \infty$ in the approximate problem (72), we need some estimates on u_j and γ_j. We provide them in the following lemmas:

Lemma 5 *The family $(u_j)_j$ is relatively compact in $C([0, T]; L^2(I))$.*

Lemma 6 *The family $(\gamma_j)_j$ is weakly relatively compact in $L^2(0, T; L^2(I))$.*

Assume that Lemmas 5 and 6 hold true. Then, we can extract from $(u_j, \gamma_j)_j$ a subsequence $(u_{j'}, \gamma_{j'})_{j'}$ such that

$$u_{j'} \to u_\infty \text{ in } C([0, T]; L^2(I)) \quad \text{and} \quad \gamma_{j'} \rightharpoonup \gamma_\infty \text{ in } L^2(0, T; L^2(I)).$$

Then we prove that

Lemma 7 *The functions u_∞ and γ_∞ satisfy the following formula*

$$\forall t \in [0, T], \quad u_\infty(t) = e^{tA}u_0(0) + \int_0^t e^{(t-s)A}\gamma_\infty(s)\, ds.$$

Moreover,

$$u_\infty \in H^1(0, T; L^2(I)) \cap L^2(0, T; D(A)) \cap C([-\tau, T]; V),$$

and

$$\forall s \in [-\tau, 0], \quad u_\infty(s) = u_0(s).$$

Finally, it remains to prove that u_∞ is solution of the original Budyko problem, and so we have to prove that γ_∞ satisfies the set inclusion. This is the object of the last lemma:

Lemma 8 *The limit function γ_∞ satisfies the set inclusion:*

$$\gamma_\infty(t)(x) + (a + bu_\infty(t, x)) - f\left(\int_{-\tau}^0 k(s, x)u_\infty(t + s, x)\, ds\right)$$
$$\in r(t)q(x)\beta(u_\infty(t, x)) \ a.e.(t, x) \in (0, T) \times I.$$
$$(75)$$

Therefore the function u_∞ is solution of the original Budyko problem (32).

Finally, note that using the Cantor diagonal process, we can extract subsequences that converge in the same way in all compact subsets of $[0, +\infty)$, hence u_∞ is well defined on $[0, +\infty)$.

9.2 Proof of Lemma 5

First note that from Theorem 6, we already know that the family $(u_j)_j$ belongs to $L^\infty([0, +\infty) \times (-1, 1))$ and is uniformly bounded in this space.

To prove that the sequence $(u_j)_j$ is relatively compact in $C([0, T]; L^2(I))$, we are going to apply the Ascoli–Arzela theorem (we refer, e.g., to [43, Theorem 1.3.1, p. 10]): we have to prove that

- $(u_j)_j$ is equicontinuous, that is:

$$\sup_j \sup_{[0,T]} \|u_j(t + h) - u_j(t)\|_{L^2(I)} \to 0 \quad \text{as } h \to 0,$$

- and that, for all $t \in [0, T]$, the set of traces $\{u_j(t), j \geq 1\}$ is relatively compact in $L^2(I)$.

The result on the set of traces $\{u_j(t), j \geq 1\}$ follows from a regularity result: we know from (73) that $u_j(t) \in V$. Moreover, it follows from the proof of Proposition 2 that there is some M^* independent of j such that

$$\sup_{t \in [0,T]} \|u_j(t)\|_V \leq M^*.$$

Since the injection of V into $L^2(I)$ is compact, we obtain that the set of traces $\{u_j(t), j \geq 1\}$ is relatively compact in $L^2(I)$.

Next, from the problem satisfied by u_j, we have the following integral representation formula

$$u_j(t) = e^{tA}u_0(0) + \int_0^t e^{(t-s)A}\gamma_j(s)\, ds,$$

hence (if $h > 0$)

$$
\begin{aligned}
u_j(t + h) - u_j(t) \\
= \left(e^{(t+h)A}u_0(0) - e^{tA}u_0(0)\right) + \left(\int_0^{t+h} e^{(t+h-s)A}\gamma_j(s)\, ds - \int_0^t e^{(t-s)A}\gamma_j(s)\, ds\right) \\
= \left((e^{(t+h)A} - e^{tA})u_0(0)\right) + \left(\int_t^{t+h} e^{(t+h-s)A}\gamma_j(s)\, ds\right) \\
+ \left(\int_0^t (e^{(t+h-s)A} - e^{(t-s)A})\gamma_j(s)\, ds\right).
\end{aligned}
$$

We estimate these three terms:

- first, by usual estimates

$$\left\|e^{(t+h)A}u_0(0) - e^{tA}u_0(0)\right\|_{L^2(I)} \leq c|h| \|u_0(0)\|_{D(A)};$$

- next,

$$\left\| \int_t^{t+h} e^{(t+h-s)A} \gamma_j(s) \, ds \right\| \le \int_t^{t+h} \|\gamma_j(s)\|_{L^2(I)} \, ds \le c|h|;$$

- finally, as we did previously,

$$\left\| \int_0^t (e^{(t+h-s)A} - e^{(t-s)A}) \gamma_j(s) \, ds \right\|_{L^2(I)} = \left\| (e^{hA} - Id) \int_0^t e^{(t-s)A} \gamma_j(s) \, ds \right\|_{L^2(I)}$$

$$= \left\| \left(\int_0^h A e^{\sigma A} \, d\sigma \right) \left(\int_0^t e^{(t-s)A} \gamma_j(s) \, ds \right) \right\|_{L^2(I)}$$

$$= \left\| \left(\int_0^h (-A)^{1/2} e^{\sigma A} \, d\sigma \right) \left(\int_0^t (-A)^{1/2} e^{(t-s)A} \gamma_j(s) \, ds \right) \right\|_{L^2(I)}$$

$$\le \left(\int_0^h \|\|(-A)^{1/2} e^{\sigma A}\|\|_{\mathcal{L}(L^2(I))} \, d\sigma \right)$$

$$\times \left(\int_0^t \|\|(-A)^{1/2} e^{(t-s)A}\|\|_{\mathcal{L}(L^2(I))} \|\gamma_j(s)\|_{L^2(I)} \, ds \right)$$

$$\le c \left(\int_0^h \frac{1}{\sqrt{\sigma}} \, d\sigma \right) \left(\int_0^t \frac{1}{\sqrt{t-s}} \, ds \right) \le c' \sqrt{h}.$$

These three estimates prove that

$$\sup_j \sup_{[0,T]} \|u_j(t+h) - u_j(t)\|_{L^2(I)} = O(\sqrt{|h|}) \quad \text{as } h \to 0.$$

Hence the family $(u_j)_j$ is equicontinuous in $C([0, T]; L^2(I))$. Therefore the Ascoli-Arzela theorem allows us to say that the family $(u_j)_j$ is relatively compact in $C([0, T]; L^2(I))$. □

9.3   Proof of Lemma 6

Since the family $(u_j)_j$ is uniformly bounded in $L^\infty((0, T), \times(-1, 1))$, we deduce that the same property holds true pour the family $(\gamma_j)_j$, hence also in the space $L^2(0, T; L^2(I))$. Hence the family $(\gamma_j)_j$ is weakly relatively compact in $L^2(0, T; L^2(I))$. □

9.4   Proof of Lemma 7

We start from the integral formula

$$u_j(t) = e^{tA} u_0(0) + \int_0^t e^{(t-s)A} \gamma_j(s) \, ds. \tag{76}$$

Choose $\xi \in L^2(I)$, and fix $t \in [0, T]$. Then

$$
\langle \xi, \int_0^t e^{(t-s)A} \gamma_j(s) \, ds \rangle_{L^2(I)} = \int_0^t \langle \xi, e^{(t-s)A} \gamma_j(s) \rangle_{L^2(I)} \, ds
$$

$$
= \int_0^t \langle e^{(t-s)A} \xi, \gamma_j(s) \rangle_{L^2(I)} \, ds = \langle z, \gamma_j \rangle_{L^2(0,T;L^2(I))}
$$

where $z \in L^2(0, T; L^2(I))$ is defined by

$$
z(s) := \begin{cases} e^{(t-s)A} \xi, & 0 \leq s \leq t, \\ 0, & t \leq s \leq T. \end{cases}
$$

Since $\gamma_{j'} \rightharpoonup \gamma_\infty$ in $L^2(0, T; L^2(I))$, we obtain that

$$
\langle z, \gamma_{j'} \rangle_{L^2(0,T;L^2(I))} \to \langle z, \gamma_\infty \rangle_{L^2(0,T;L^2(I))} \qquad \text{as } j' \to \infty.
$$

But

$$
\langle z, \gamma_\infty \rangle_{L^2(0,T;L^2(I))} = \int_0^t \langle z(s), \gamma_\infty(s) \rangle_{L^2(I)} \, ds = \int_0^t \langle e^{(t-s)A} \xi, \gamma_\infty(s) \rangle_{L^2(I)} \, ds
$$

$$
= \int_0^t \langle \xi, e^{(t-s)A} \gamma_\infty(s) \rangle_{L^2(I)} \, ds = \langle \xi, \int_0^t e^{(t-s)A} \gamma_\infty(s) \, ds \rangle_{L^2(I)}.
$$

From the fact that $u_{j'} \to u_\infty$ in $C([0, T]; L^2(I))$, we deduce from (76) that

$$
u_\infty(t) = e^{tA} u_0(0) + \int_0^t e^{(t-s)A} \gamma_\infty(s) \, ds.
$$

Finally, since $\gamma_\infty \in L^2(0, T; L^2(I))$, and $u_0(0) \in D(A)$, u_∞ has the regularity claimed in Lemma 7. $\qquad \square$

9.5 Proof of Lemma 8

It remains to identify the weak limit γ_∞. In order to do this, fix $\kappa > 0$ and let us introduce

$$
Q_\kappa := \{(t, x) \in (0, T) \times (-1, 1), |r(t)q(x)| \geq \kappa\}, \tag{77}
$$

and then we define on Q_κ the function

$$\mathcal{B}(t,x) := \frac{1}{r(t)q(x)}\left(\gamma_\infty(t,x) + (a+bu_\infty) - f\left(\int_{-\tau}^0 k(s,x)u_\infty(t+s,x)\,ds\right)\right).$$
(78)

Of course this definition is motivated by the fact that, also on Q_κ, we have

$$\beta_j(u_j(t,x)) = \frac{1}{r(t)q(x)}\left(\gamma_j(t,x) + (a+bu_j) - f\left(\int_{-\tau}^0 k(s,x)u_j(t+s,x)\,ds\right)\right).$$

Hence we immediately have that

$$\beta_{j'}(u_{j'}) \rightharpoonup \mathcal{B} \quad \text{in } L^2(Q_\kappa) \text{ as } j' \to \infty.$$

This already tells us that

$$a_i \le \mathcal{B}(t,x) \le a_f \quad a.e.\ (t,x) \in Q_\kappa.$$

Indeed, choose E any measurable part of Q_κ and χ_E its characteristic function: then

$$\int\int_E (\mathcal{B} - a_i) = \int\int_{Q_\kappa} (\mathcal{B} - a_i)\chi_E = \lim_{j'\to\infty} \int\int_{Q_\kappa} (\beta_{j'}(u_{j'}) - a_i)\chi_E.$$

Since $\beta_{j'}(u_{j'}) - a_i \ge 0$, the last quantity is nonnegative, and hence

$$\int\int_E (\mathcal{B} - a_i) \ge 0.$$

Since this holds true for all E, we obtain that $\mathcal{B} - a_i \ge 0$ on Q_κ. In the same way, $\mathcal{B} - a_f \le 0$ on Q_κ.

Now we conclude by proving that if

$$(t,x) \in \tilde{Q}_\kappa, \quad \text{and} \quad u_\infty(t,x) < \bar{u},$$

then

$$\mathcal{B}(t,x) = a_i.$$

For $\eta > 0$, we introduce

$$D_{\kappa,\eta} := \{(t,x) \in Q_\kappa \mid u_\infty(t,x) \le \bar{u} - \eta\}.$$

Up to some subsequence, we can assume that

$$u_{j'} \to u_\infty \quad a.e. \ (t, x) \in (0, T) \times (-1, 1).$$

If $(t, x) \in D_{\kappa, \eta}$, there exists $j'(t, x)$ large enough such that

$$\forall j' \geq j'(t, x), \quad u_{j'}(t, x) \leq \bar{u} - \frac{\eta}{2}.$$

And then, the construction of $\beta_{j'}$ implies that, if j' is large enough, we have

$$\beta_{j'}(u_{j'}(t, x)) = a_i.$$

Hence

$$\beta_{j'}(u_{j'}) \to a_i \quad a.e. \ (t, x) \in D_{\kappa, \eta}.$$

Now, consider $\psi \in L^2(Q_\kappa)$. We have

$$\int \int_{Q_\kappa} (\beta_{j'}(u_{j'}) - \mathcal{B})\psi \to 0 \quad \text{as } j' \to +\infty.$$

Moreover, if ψ is supported in $D_{\kappa, \eta}$, then, since $\beta_{j'}(u_{j'}) \to a_i$ a.e. $D_{\kappa, \eta}$, we deduce from the using the dominated pointwise convergence theorem that

$$\int \int_{Q_\kappa} (\beta_{j'}(u_{j'}) - \mathcal{B})\psi \to \int \int_{Q_\kappa} (a_i - \mathcal{B})\psi \quad \text{as } j' \to +\infty.$$

Hence

$$\int \int_{Q_\kappa} (a_i - \mathcal{B})\psi = 0$$

and this holds true for all ψ supported in $D_{\kappa, \eta}$, therefore

$$a_i = \mathcal{B} \text{ on } D_{\kappa, \eta}.$$

Letting $\eta \to 0$, we obtain that

$$\mathcal{B} = a_i \quad \text{on } \{(t, x) \in Q_\kappa \mid u_\infty(t, x) < \bar{u}\}.$$

We can proceed in the same way to prove that

$$\mathcal{B} = a_f \quad \text{on } \{(t, x) \in Q_\kappa \mid u_\infty(t, x) > \bar{u}\}.$$

And finally, letting $\kappa \to 0$, we obtain that, if $r(t)q(x) \neq 0$, we have

$$\frac{1}{r(t)q(x)}\left(\gamma_\infty(t,x) + (a + bu_\infty) - f\left(\int_{-\tau}^{0} k(s,x)u_\infty(t+s,x)\,ds\right)\right)$$
$$\begin{cases} = a_i & \text{if } u_\infty(t,x) < \bar{u}, \\ = a_f & \text{if } u_\infty(t,x) > \bar{u}, \\ \in [a_i, a_f] & \text{in any case,} \end{cases}$$

Of course, if $r(t)q(x) = 0$, then

$$\gamma_{j'}(t,x) = -(a + bu_{j'}(t,x)) + f(H_{j'}),$$

and we deduce that

$$\gamma_\infty(t,x) = -(a + bu_\infty(t,x)) + f\left(\int_{-\tau}^{0} k(s,x)u_\infty(t+s,x)\,ds\right).$$

Hence

$$\gamma_\infty(t)(x) = r(t)q(x)\mathcal{B}(t,x) - (a + bu_\infty(t,x)) + f(H_\infty(t,x))$$
$$\in r(t)q(x)\beta(u_\infty(t,x)) - (a + bu_\infty(t,x)) + f(H_\infty(t,x))$$

a.e. $(t,x) \in (0,T) \times I$, hence the set inclusion (75) is satisfied. $\square$

Acknowledgements This research was partly supported by Istituto Nazionale di Alta Matematica through the European Research Group GDRE CONEDP. The authors acknowledge the MIUR Excellence Department Project awarded to the Department of Mathematics, University of Rome Tor Vergata, CUP E83C18000100006. The authors also wish to thank K. Fraedrich for a very interesting discussion on the question of Energy Balance Models with Memory, and M. Ghil for inspiring remarks.

References

1. Bensoussan, A., Da Prato, G., Delfour, M.C., Mitter, S.K.: Representation and control of infinite-dimensional systems. In: Systems and Control: Foundations and Applications, vol. 1. Birkhäuser, Boston (1992)
2. Bhattacharya, K., Ghil, M., Vulis, I.L.: Internal variability of an energy-balance model with delayed albedo effects. J. Atmos. Sci. **39**, 1747–1773 (1982)
3. Bódai, T., Lucarini, V., Lunkeit, F., Boschi, R.: Global instability in the Ghil-Sellers model. Clim. Dyn. **44**, 3361–3381 (2015)
4. Budyko, M.I.: The effect of solar radiation variations on the climate of the Earth. Tellus **21**(5), 611–619 (1969)
5. Campiti, M., Metafune, G., Pallara, D.: Degenerate self-adjoint evolution equations on the unit interval. Semigroup Forum **57**, 1–36 (1998)
6. Cannarsa, P., Martinez, P., Vancostenoble, J.: Null controllability of degenerate heat equations. Adv. Differ. Equ. **10**(2), 153–190 (2005)

7. Cannarsa, P., Martinez, P., Vancostenoble, J.: Carleman estimates for a class of degenerate parabolic operators. SIAM J. Control Optim. **47**(1), 1–19 (2008)
8. Cannarsa, P., Rocchetti, D., Vancostenoble, J.: Generation of analytic semi-groups in L^2 for a class of second order degenerate elliptic operators. Control Cybern. **37**(4), 831–878 (2008)
9. Cannarsa, P., Martinez, P., Vancostenoble, J.: Global Carleman Estimates for Degenerate Parabolic Operators with Applications. Memoirs of the American Mathematical Society, vol. 239, no. 1133. American Mathematical Society, Providence, RI (2016)
10. Cannarsa, P., Floridia, G., Khapalov, A.Y.: Multiplicative controllability for semilinear reaction-diffusion equations with finitely many changes of sign. J. Math. Pures Appl. **108**(4), 425–458 (2017)
11. Cristofol, M., Roques, L.: Stable estimation of two coefficients in a nonlinear Fisher-KPP equation. Inverse Prob. **29**(9), 095007, 18 pp. (2013)
12. Diaz, J.I.: Mathematical analysis of some diffusive energy balance models in climatology. In: Diaz, J.I., Lions, J.L. (eds.) Mathematics, Climate and Environment, pp. 28–56. Masson, Paris (1993)
13. Diaz, J.I.: On the mathematical treatment of energy balance climate models. In: The Mathematics of Models for Climatology and Environment. NATO ASI Series. Series I: Global Environmental Change, vol. 48. Springer, Berlin (1997)
14. Diaz, J.I.: Diffusive energy balance models in climatology. In: Studies in Mathematics and its Applications, vol. 31. North-Holland, Amsterdam (2002)
15. Diaz, J.I., Hetzer, G.: A functional quasilinear reaction-diffusion equation arising in climatology. In: Équations aux dérivées partielle et applications. Articles dédiés à J.L. Lions, pp. 461–480. Elsevier, Paris (1998)
16. Diaz, J.I., Tello, L.: A nonlinear parabolic problem on a Riemannian manifold without boundary arising in climatology. Collect. Math. **50**, 19–51 (1997)
17. Diaz, J.I., Hetzer, G., Tello, L.: An energy balance climate model with hysteresis. Nonlinear Anal. **64**, 2053–2074 (2006)
18. Floridia, G.: Approximate controllability of nonlinear degenerate parabolic equations governed by bilinear control. J. Differ. Equ. **257**(9), 3382–3422 (2014)
19. Floridia, G., Nitsch, C., Trombetti, C.: Multiplicative controllability for nonlinear degenerate parabolic equations between sign-changing states. ESAIM COCV. https://doi.org/10.1051/cocv/2019066
20. Fraedrich, K.: Structural and stochastic analysis of a zero-dimensional climate system. Q. J. R. Meteorol. Soc. **104**, 461–474 (1978)
21. Fraedrich, K.: Catastrophes and resilience of a zero-dimensional climate system with ice-albedo and greenhouse feedback. Q. J. R. Meteorol. Soc. **105**, 147–167 (1979)
22. Ghil, M.: Climate stability for a Sellers-type model. J. Atmos. Sci. **33**, 3–20 (1976)
23. Ghil, M., Childress, S.: Topics in Geophysical Fluid Dynamics: Atmosphere Dynamics, Dynamo Theory, and Climate Dynamics. Springer, New York (1987)
24. Guerrero, S., Imanuvilov, O.Y.: Remarks on non controllability of the heat equation with memory. ESAIM COCV **19**, 288–300 (2013)
25. Held, I.M., Suarez, M.J.: Simple albedo feedback models of the ice caps. Tellus **26**, 613–629 (1974)
26. Hetzer, G.: Global existence, uniqueness, and continuous dependence for a reaction-diffusion equation with memory. Electron. J. Diff. Equ. **5**, 1–16 (1996)
27. Hetzer, G.: The number of stationary solutions for a one-dimensional Budyko-type climate model. Nonlinear Anal. Real World Appl. **2**, 259–272 (2001)
28. Hetzer, G.: Global existence for a functional reaction-diffusion problem from climate modeling. Discrete Contin. Dyn. Syst. **31**, 660–671 (2011)
29. Imanuvilov, O.Y., Yamamoto, M.: Lipschitz stability in inverse parabolic problems by the Carleman estimates. Inverse Prob. **14**(5), 1229–1245 (1998)
30. Lenton, T.M., Held, H., Kriegler, E., Hall, J.W., Lucht, W., Rahmstorf, S., Schellnhuber, H.J.: Tipping elements in the Earth's climate system. Proc. Nath. Acad. Sci. U. S. A. **105**, 1786–1793 (2008)

31. Lions, J.L., Magenes, E.: Problèmes aux limites non homogènes et applications, vol. 1 (French). Travaux et recherches mathématiques, No. 17. Dunod, Paris (1968)
32. Martinez, P., Tort, J., Vancostenoble, J.: Lipschitz stability for an inverse problem for the 2D-Sellers model on a manifold. Riv. Mat. Univ. Parma **7**, 351–389 (2016)
33. North, G.R., Mengel, J.G., Short, D.A.: Simple energy balance model resolving the season and continents: applications to astronomical theory of ice ages. J. Geophys. Res. **88**, 6576–6586 (1983)
34. Pandolfi, L.: Riesz systems, spectral controllability and a source identification problems for heat equation with memory. Discrete Contin. Dyn. Syst. **4**(3), 745–759 (2011)
35. Pazy, A.: Semigroups of Linear Operators and Applications to Partial Differential Equations. Applied Mathematical Sciences. Springer, New York (1983)
36. Protter, M.H., Weinberger, H.F.: Maximum Principles in Differential Equations. Springer, New York (2012)
37. Roques, L., Cristofol, M.: On the determination of the nonlinearity from localized measurements in a reaction-diffusion equation. Nonlinearity **23**(3), 675–686 (2010)
38. Roques, L., Checkroun, M.D., Cristofol, M., Soubeyrand, S., Ghil, M.: Determination and estimation of parameters in energy balance models with memory. Proc. R. Soc. A **470**, 20140349 (2014)
39. Sellers, W.D.: A global climatic model based on the energy balance of the earth-atmosphere system. J. Appl. Meteorol. **8**(3), 392–400 (1969)
40. Tanabe, H.: On the equations of evolution in a Banach space. Osaka Math. J. **12**(2), 363–376 (1960)
41. Tao, Q., Bao, H.: On the null controllability of heat equation with memory. J. Math. Anal. Appl. **440**, 1–13 (2016)
42. Tort, J., Vancostenoble, J.: Determination of the insolation function in the nonlinear climates Sellers model. Ann. I. H. Poincaré **29**, 683–713 (2012)
43. Vrabie, I.I.: Compactness Methods for Nonlinear Evolutions. Pitman Monographs and Surveys in Pure and Applied Mathematics. Longman Scientific & Technical, Essex (1987)
44. Walsh, J., Rackauckas, C.: On the Budyko-Sellers energy balance energy climate model with ice line coupling. Discrete Contin. Dyn. Syst. B **20**, 2187–2216 (2015)
45. Zaliapin, I., Ghil, M.: Another look at climate sensitivity. Nonlinear Process. Geophys. **17**, 113–122 (2010)

Part II
Theme: Hydrology

Bedload Transport Processes in a Coastal Sand-Bed River: The Study Case of Fiumi Uniti River in the Northern Adriatic

Silvia Cilli, Paolo Billi, Leonardo Schippa, Edoardo Grottoli, and Paolo Ciavola

Abstract Over the last decades, most of the Emilia-Romagna (Italy) beaches have been affected by marked erosion that is still progressing, which is primarily due to the reduction of sediment supply by the local rivers. In addition to larger fluvial systems, the role of small rivers has been recognized as important in contributing to both beach stability and changes. Bedload measurements were carried out by previous authors in 2005–2006 but, in order to widen the data set of river sediment supply for the whole region, a new bedload measurement campaign was performed in 2017 on the Fiumi Uniti River, a typical example of a small river system whose sediment supply variations substantially influence the stability of the beaches adjoining its mouth. Marked changes were observed in the bedload transport rates and bed material grain size between the two sampling campaigns of 2005/2006 and 2017. The threshold conditions for bed particle entrainment are analysed by means of the Shields classical criterion and other approaches. The results indicate that all these criteria are unable to predict the actual value, except for Carling and Bagnold criteria. This result has large implications in the assessment of bedload yield in sand-bed rivers.

Keywords Bedload transport · Flood measurements · Threshold condition · Critical shear stress

S. Cilli (✉) · E. Grottoli · P. Ciavola
Department of Physics and Earth Sciences, University of Ferrara, Ferrara, Italy
e-mail: silvia.cilli@unife.it; edoardo.grottoli@unife.it; paolo.ciavola@unife.it

P. Billi
IPDRE, Tottori University, Tottori, Japan
e-mail: Paolo.billi@alrc.tottori-u.ac.jp

L. Schippa
Department of Engineering, University of Ferrara, Ferrara, Italy
e-mail: leonardo.schippa@unife.it

© Springer Nature Switzerland AG 2020
P. Cannarsa et al. (eds.), *Mathematical Approach to Climate Change and Its Impacts*, Springer INdAM Series 38,
https://doi.org/10.1007/978-3-030-38669-6_3

133

1 Introduction

Rivers are the main natural network for the sediment transfer to beaches, contributing to their formation and stability. The amount of sediment that is released at their mouths is the result of complex hydrological and hydraulic phenomena. Yet, it depends on the geomorphological dynamics of the river basin, but it is also influenced by human activities and infrastructures, which, in a direct or indirect way, control the evolution of the catchment as well as the marine-coastal area. Anthropogenic impacts, such as changes in land use (mainly increase in the forested land), river bed mining, prescnce of engineering works and progressive dismantling of the rivers mouths, are only some of them [2, 13, 17, 20, 21]. For all these reasons, in the last decades, many beaches, particularly in Italy and other industrialized countries, have been affected by marked erosion that is still progressing. The Emilia-Romagna is an Italian region particularly affected by this phenomenon, which is mainly due to the reduction of sediment supply by the local rivers. As well as for larger fluvial systems, the role of small rivers has been recognized as important in contributing to both beach stability and changes [18]. Unfortunately, despite few sporadic field studies, bed-load sediment transport in the Emilia-Romagna region is poorly known and referred to a limited number of rivers [5, 6, 9, 10]. In order to widen the data set of river sediment supply in the whole region, bedload measurement campaigns in representative rivers are in progress by the authors of the current paper. As part of a regional scale project, which includes hydrological investigation and bed load transport monitoring, the Fiumi Uniti study is a pilot study aiming to define the sediment supply to the whole Romagna coast. Main points of this project are a review of existing bedload field data and analysis of additional data in order to qualify and quantify the sediment budget of the Fiumi Uniti river, a typical example of a small river system in the Emilia Romagna region. The research activities include hydrological investigations and bed load transport measurements about 8 km upstream of the river mouth.

2 Study Area

The Fiumi Uniti catchment is about $1000 \, \text{km}^2$ and derives from the unification of two rivers: the Montone and Ronco. Both rivers have similar morphometric characteristics (Table 1). They flow from the Apennine Mountains to the Padan plain, where they merge near Ravenna. The river mouth is located between Lido Adriano e Lido di Dante, on the Adriatic Sea (Fig. 1). The Fiumi Uniti is an artificial river 9.3 km long (from the confluence to the sea) that was realized by the citizen of Ravenna around the 1700s in order to connect the Montone and Ronco rivers to the Adriatic Sea to avoid repeated floods in the city.

Two-thirds of the basin surface are located in the northern Apennines (maximum elevation 1650 m) and are underlain by Miocene turbidites consisting of sandstones

Table 1 Morphometric
parameters of the Montone
and Ronco rivers

Parameters	Units	Montone	Ronco
Area	km^2	441.09	524.32
Perimeter	km	600.53	545.96
H_{max}	m asl	1245.17	1649.52
$H_{closure}$	m asl	30.03	24.84
H_{mean}	m asl	450.36	484.94
S_{mean}	%	39.06	39.38
$L_{main\,reach}$	km	90.19	92.65
R_c	–	0.008	0.011
R_f		0.34	0.38
R_{el}	–	0.26	0.28
C_{comp}	–	25.33	21.11
F_f	–	0.05	0.06
LPD	km	58.00	68.89
DD	km km^{-2}	0.13	0.13

The listed parameters are: area, perimeter, maximum altitude (H_{max}), altitude at the closure of
the basin ($H_{closure}$), mean altitude (H_{mean}), mean
slope of the basin (S_{mean}); circularity ratio (R_c),
basin shape ratio (R_f), elongation ratio (R_{el}),
compactness coefficient (C_{comp}), form factor
(F_f), longest drainage path (LPD) and drainage
density (DD)

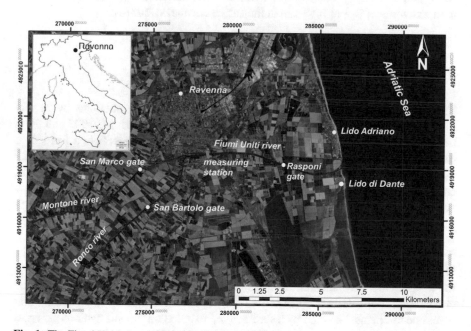

Fig. 1 The Fiumi Uniti river with measuring site and sluice gates

and marlstones alternations [1]. The lower portions of the catchments are underlain by Pliocene marine deposits and the Quaternary Po river alluvial deposits [1]. The climate is subcontinental temperate [22] with a mean temperature averaged over the 1961–2010 period of 13–13.5 °C for most of the basin and a mean annual cumulated precipitation, averaged over the same period, that varies from 600 mm in the coastal areas to 1800 mm (see Fig. 2) in the catchment headwaters [3].

The Fiumi Uniti river is a typical example of a small river system, which is regulated by a sluice gate dam, located 3.5 km upstream of the river outlet. The sluice gate is closed in summer to prevent saltwater intrusion. However, during the other seasons and during floods, even the smallest ones, the sluice gate is kept open, allowing the river sediment to reach the beach. Recently, the local Water Authority decided to keep the sluice gate permanently open. At low flow the tidal influence is clearly observed in the data, though the tide range at the gate is evident, even though limited to about 0.4 m.

The Montone and Ronco rivers have sluice gates too, located a few kilometres upstream the river junction. These gates are for irrigation purpose, and they are regulated according to the seasonal rainfall conditions: they are open during the flood season (October-March) and closed during the dry ones (April-September). The monitoring station is located in Ravenna (Fig. 1) at a pedestrian bridge.

In this reach the river shows a prismatic channel (Fig. 3) having a rectangular cross-section (Fig. 3) and a streambed gradient of 0.00029 m/m [6]. The maximum channel width at the levee crest level is about 60 m and the active channel width, i.e. where bed load transport occurs, is 40 m. The bed material is predominantly sandy, with an averaged D_{50} of 0.43 mm in the active channel width [6].

3   Methods and Instrumentation

The analysis of sediment transport is based only on data obtained by in situ measurements at the monitoring station. Part of them comes from 2005 and 2006 field measurements by Billi et al. [6]. Additional data were obtained from field measurement during two floods occurred on 07/02/2017 and 07/03/2017. During these floods, hydraulic and sediment monitoring was carried out at five verticals, equally spaced across the active channel width. At each vertical flow depth, flow velocity and bedload transport were measured. Flow velocity was measured with a standard USGS AA type current meter, with vertical axis. Bedload transport was sampled with a standard Helley-Smith bedload sampler (US BL-84) with a 76×76 mm intake and an expansion rate of 1.10, which is considered to provide the highest efficiency [12]. The sample bag of the Helley Smith had a 0.1 mm mesh. All the instruments were lowered from the pedestrian bridge with the help of a wheel crane. Measurements were taken at established time intervals according to the water level rate of change detected by a staff gauge, installed on the left bank of the cross section. The measurements across the whole cross-section required about 1 h. As pointed out by Emmet [12], the sampling time of the Helley-Smith sampler is not

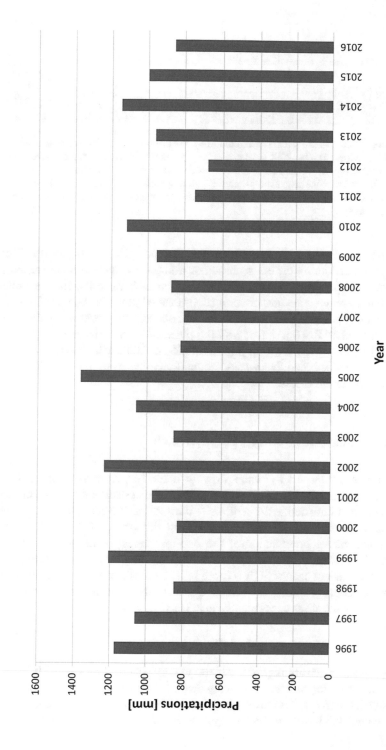

Fig. 2 Annual precipitation on the Fiumi Uniti river basin in the 1996–2016 interval, data obtained from the Italian Hydrographic Service

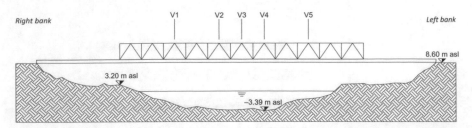

Fig. 3 The cross-section of the river. The figure shows the position of the five verticals used for the monitoring activities. The bankfull level can be approximately fixed around 3 m above mean sea level, i.e. at the level of the first bank (3.20 m asl), as illustrated. The figure shows also the altitude of the stream bed at the present cross section (thalweg equal to −3.39 m asl) and the one of the main bank (8.60 m asl). The values reported are refereed to the cross section condition before the monitoring activities

standardized and has to be calibrated after a few attempts, given the high variability of bedload transport during different phases of a flood. All the sediment samples collected were cleared from vegetation debris and organic material. Each sample was wet sieved to remove the fine fraction (finer than 63 µm). The coarser fraction was dried for 1 day at a temperature of 105 °C. Then each sample was dry-sieved for 20 min with a standard Ro-Tap shaker with 0.5 phi scale sieves. The sample portion finer than 63 µm was not included in the calculation of bedload since that material is part of the suspended load that was accidentally collected by the sampler bag, together with vegetation debris.

4 Mathematical and Statistical Approach

Bed-load sediment transport by water flow in natural streams depends on bed material characteristics and flow condition. Usually, classical bed-load formulas available in the literature define the sediments transport as the continuous contacts of the particles with the bed strictly limited by the effect of the gravity. In fact the mathematical representation of the bed-load transport motion is mainly distinguished in: rolling, sliding and saltation. The reach of study shows a regular geometry, and the rate of change of flow (both in terms of water level and discharge) is relatively weak during a flood, therefore a quasi steady gradually varied flow has been assumed; and the shear stress results as in Eq. (1):

$$\tau = \rho g R S \qquad (1)$$

where: ρ is the water density [kg/m^3], g is the gravity acceleration [m/s^2], R is the hydraulic radius [m], for wide and relatively shallow rivers R can be substituted by the mean flow depth H, and S is the energy gradient slope [m/m]. In the present study it has been calculated in two different ways: in the first case, it has been assumed equal to the stream bed slope, while in the second one it has been simulated through the Hec-Ras model.

When the value of the bed shear velocity exceeds the threshold value for the initiation of motion, the particles will start moving constantly in contact with the bed. In the present analysis the transport is defined as previously explained. The criterion here used to predict the initiation of sediment motion is the Shields classical criterion [24] which is the most used in river dynamics and fluvial geomorphology. It introduces a dimensionless number of the critical shear stress θ_{cr} as shown in Eq. (2):

$$\theta_{cr} = \tau_{cr}/(\rho_s - \rho)gD, \tag{2}$$

where: τ_{cr} is the shear stress corresponding to the incipient motion [N/m^2], ρ_s is the density of the sediment [kg/m^3], ρ is the density of water [kg/m^3], g is the gravity acceleration [m/s^2], and D is the characteristic particle diameter of bed material [m]. Generally the D_{50}, which is the median value of the particle size distribution, is used as the characteristic particle diameter of the sediment.

The Shields curve corresponding to the threshold condition has been represented in analytic form by Brownlie [7] as shown in Eq. (3):

$$\theta_{cr} = 0.22(R_p)^{-0.6} + 0.06exp(-7.7R_p^{-0.6}), \tag{3}$$

where R_p is equal to $(R^*((\rho_s/\rho)-1))^{0.5}$, R^* is the dimensionless Reynolds number defined as $(u_* D)/v$, $u_* = \sqrt{\tau/\rho}$ [m/s] is the bed shear velocity, D the characteristic particle diameter of the sediment [m] and v the kinematic viscosity [m^2/s].

5 Results and Discussion

Eleven floods were monitored from 12 April 2005 to 7 March 2017. In this study only nine of them are considered, i.e. those monitored with the sluice gates fully open. The weakest flood occurred on 11/10/2005 with a flow discharge of 17.27 m^3/s, whereas the largest one occurred on 12/04/2005 with 358.16 m^3/s, which can be considered one of the largest floods recorded in the last decade [6]. All the floods monitored are reported in Fig. 4, where different flow rating curves are also shown. The data may be represented by the flow rating curve proposed by Herschy [15] with the form $Q = a(x - b)^c$, in which x is water depth and a, b and c constants. The rating curve is expressed by the dark grey solid line in Fig. 4 and by the linear regression (the black dotted line). Nevertheless, the power function (light gray solid line) and the polynomial regression (grey dashed line) have the best fitting (Fig. 4). In fact, the RMSE (Root Mean Square Error) estimated is 126.46, 127.47, 124.38 and 121.88, respectively.

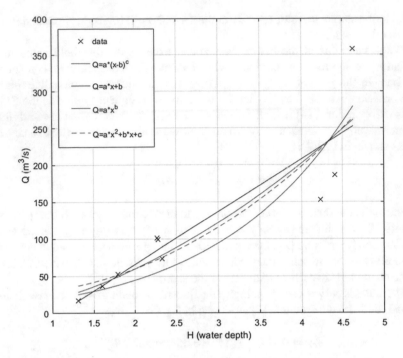

Fig. 4 Rating curves and summary of measured data. Correlation between water discharge (Q) and water depth (H)

For each flood, bedload discharge, Q_b, has been calculated in t/day (Fig. 5). The field observations indicate that bedload was active across the entire cross-section only for big flow discharges, whereas for smaller floods, only the central portion of the river bed was involved. For the first seven floods, the D_{50} varied between 0.296 and 0.626 mm with a mean of 0.43 mm, whereas for the last two floods D_{50} varied from 0.34 to 0.97 mm, with a mean of 0.55 mm. In light of the data collected, it is not definitively possible to establish the functional relationship between liquid and solid discharge. Figure 5 shows the trend of the monomial function with exponent 1 and 1.24, respectively. The best correlation is expressed by the linear function (black dashed line), with a RMSE equal to 231 (Fig. 5).

It is worth noticing that the 2017 data, indicated with the black circle in Fig. 5, correspond to low values of bedload transport despite the flow discharges were relatively high. This result could be accounted for the likely permanent opening of the Fiumi Uniti sluice gate which influenced the sediment movement. With the sluice gate opened there is no chance for upstream sediment accumulation. By contrast, during the 2005–2006 measurements, when the sluice gate was closed during low flows, even low discharges were able to remove the sediment accumulated during the closing period of the sluice gate. In order to confirm this hypothesis, the bed shear stress relative to each flow discharge was calculated and

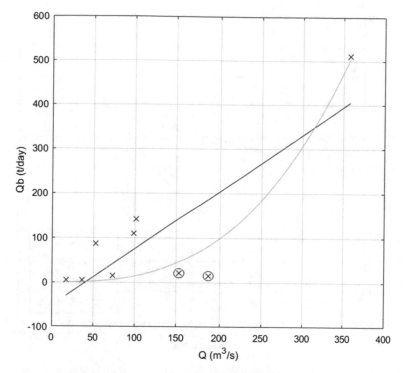

Fig. 5 Correlation between flow discharge (Q) and sediment discharge (Qb). The circled data refer to the measurements of 2017

it is shown in Fig. 6. In this case, the linear regression (black solid line, in Fig. 6) demonstrates the best fitting (RMSE equal to 2.92).

Unfortunately, very few studies have investigated the threshold conditions for sandy bed material entrainment, but from our field measurements it is nearly possible to identify the critical flow for the smallest, appreciable bedload transport. This result is compared with the predictions of a few of the most used criteria available in the literature for rivers with sandy or gravelly sand bed material. By applying the classical Shield's criterion, as it is represented by Brownlie [7] and herein reported in Eq. (1), to the threshold flow observed in the field ($17.27 \, \text{m}^3/\text{s}$ and $\tau_c = 3.73 \, \text{N/m}^2$) we have the following values of the critical shear stress: $0.37 \, \text{N/m}^2$ (considering $D_{50} = 0.43 \, \text{mm}$) and $0.47 \, \text{N/m}^2$ (considering $D_{50} = 0.55 \, \text{mm}$), respectively (Table 2). The resulting Shield's dimensionless parameter, θ_{cr}, resulted 0.050 (both for $D_{50} = 0.43 \, \text{mm}$ and $D_{50} = 0.55 \, \text{mm}$). These values are lower than the typical $\theta_{cr} = 0.056$, values commonly used in bedload transport formulae [16, 19]. Along with Shileds-Brownlie criterion for the incipient motion, several other approaches have been considered. These criteria follow an empirical power law of the type $\tau_{cr} = aD^b$, where a and b are two fitting parameters determined by experimental data analysis. One of the first formulas proposed in the literature are that of Carling [8] where $\tau_{cr} = 6.33D^{0.38}$ and that of Costa [11] where

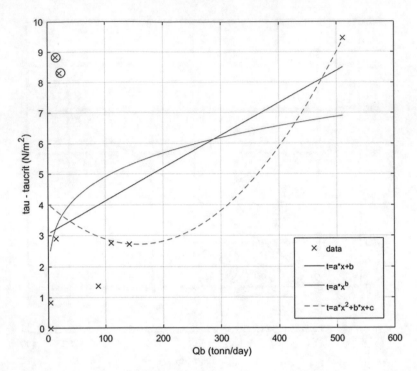

Fig. 6 Correlation between sediment discharge (Qb) and relative shear stress (τ). The circled data refer to the measurements of 2017

$\tau_{cr} = 26.6D^{1.21}$. Hammond et al. [14] proposed $\tau_{cr} = 55D^{0.42}$. Moreover the critical stream power approach, originally represented by Bagnold [4] and then revised by Parker et al. [23], has been considered as well in Eq. (4):

$$\omega_c = 2860.5D^{1.5}\log(12H/D) \qquad [kg/(m\ s)], \qquad (4)$$

where: ω_c represents the threshold value for the stream power, D is the characteristic particle diameter of the sediment [m], and H is the mean water depth [m].

In this case, the critical shear stress was obtained by the following expression (5):

$$\tau_{cr} = g(\omega_c/v) \qquad [N/m^2], \qquad (5)$$

in which: v is the mean flow velocity [m/s], and g is the gravity acceleration, equal to 9.81 m/s^2.

All these criteria for the threshold condition were tested against the field data, considering again both 2005/2006 and 2017 bed material D_{50} (Table 2).

Table 2 shows that Carling [8] and Bagnold [4] provide the best predictions. In fact their results are very close to the actual value obtained from the field data

Table 2 Comparison of critical shear stress values obtained by different criteria

	τ_{cr} (D50 = 0.00043 m) [N/m^2]	τ_{cr} (D50 = 0.00055 m) [N/m^2]
Brownlie [7]	0.37	0.47
Carling [8]	3.33	3.65
Costa [11]	0.06	0.08
Hammond et al. [14]	1.47	1.63
Bagnold [4]	3.05	4.31

(3.73 N/m^2). Conversely, the equation of Costa [11] under predicts critical shear stress and such a large difference can be accounted for by the larger grain size (gravel) of bed material on which it is based. Hammond et al. equation is derived for the finer grain size (5 mm), and though it was based on data from tidal estuaries [14], its results are not satisfactory. The results of Table 2 clearly indicated that, with the exception of Bagnold and Carling, almost all the criteria considered are not reliable to predict critical shear stress in a sand bed costal river like the Fiumi Uniti. These criteria, in fact, are mainly based on gravel and fine gravel-sand mixture and therefore did not take into account the roughness component of bedforms (e.g. dunes) that are present in the study reach. This issue will be matter of future studies. Moreover, the critical shear stress has been calculated also by the Hec-Ras model, using as input flow the threshold discharge observed in the field (i.e., 17.27 m^3/s). The energy slope obtained from the model is equal to 0.000135 m/m, therefore the τ_{cr} calculated by Eq. (1) resulted equal to 1.73 N/m^2. The value estimated by Hec-Ras is substantially lower than that calculated with field data (3.73 N/m^2).

6 Conclusions

The present study, as part of a regional scale project, aims to widen the data set of river sediment supply to the beaches of the whole Emilia Romagna region, which in the last decades were characterized by a marked erosion phenomenon. Bedload measurement campaigns in the Fiumi Uniti River, considered as a representative river of the region, were carried out and are still in progress.

The field measurements already performed allowed to obtain significant data related to the sediment size of the bedload, as well as to quantify bedload transport for each flood in association with flow discharge. Finally, the critical shear stress for the sandy bed material entrainment was investigated. The data evidenced a difference in bedload transport between the 2005 and 2006 field campaigns and a new data-set of field measurements undertaken in 2017. In particular, 2017 data demonstrated a decrease in the sediment transport rate, probably due to a difference flow regime at the presence downstream sluice gate. Although the sluice gate was previously maintained opened even during small floods, in recent years it has been kept permanently open. This fact has to be investigated in more detail in order to explain the increase in bed material median grain size recorded in 2017. A comparison with the results of well-known criteria to predict the threshold conditions for bed particle entrainment indicates that these criteria largely under predict the value of critical shear stress, whereas the classical Bagnold criterion slightly over predicts the actual threshold. In this context the Carling criterion [8] seems to be the most acceptable one. Further studies are needed, especially to incorporate the roughness effect of moving dune bedforms.

References

1. Amorosi, A., Centineo, M.C., Dinelli, E., Lucchini, F., Tateo, F.: Geochemical and mineralogical variations as indicators of provenance changes in Late Quaternary deposits of SE Po Plain. Sediment. Geol. **151**, 273–292 (2002)
2. Anthony, E.J.: The Human influence on the Mediterranean coast over the last 200 years: a brief appraisal from a geomorphological perspective. Geomorphologie **20**(3), 219–226 (2014)
3. Antolini, G., Auteri, L., Pavan, V., Tomei, F., Tomozeiu, R., Marletto, V.: A daily high-resolution gridded climatic data set for Emilia-Romagna, Italy, during 1961–2010. Int. J. Climatol. **36**, 1970–1986 (2016)
4. Bagnold, R.A.: An empirical correlation of bedload transport rates in flumes and natural rivers. Proc. R. Soc. Lond. **A372**, 453–473 (1980)
5. Billi, P., Salemi, E.: Misura delle portate solide in sospensione al fondo del F. Reno. Arpa Rivista, supplemento al n. 6, 8–15 (2004)
6. Billi, P., Salemi, E., Preciso, E., Ciavola, P., Armaroli, C.: Field measurement of bedload in a sand-bed river supplying a sediment starving beach. Z. Geomorphol. **61**(3), 207–223 (2017)
7. Brownlie, W.R.: Prediction of flow depth and sediment discharge in open channels. Report No. KH-R-43A, p. 232. California Institute of Technology, W. M. Keck Laboratory, Pasedena (1981)
8. Carling, P.A.: Threshold of coarse sediment transport in broad an narrow natural streams. Earth Surf. Process. Landf. **8**, 1–18 (1983)
9. Ciavola, P., Billi, P., Armaroli, C., Preciso, E., Salemi, E., Balouin, Y.: Morphodynamics of the Bevano Stream outlet: the role of bedload yield. Valutazione della morfodinamica di foce del Torrente Bevano (RA): il ruolo del trasporto solido di fondo. Geologia Tecnica e Ambientale **1**, 41–57 (2005)
10. Ciavola, P., Salemi, E., Billi, P.: Sediment supply and morphological evolution of a small river mouth (Fiumi Uniti, Ravenna, Italy): should river management be storm-driven? In: Proceedings of the China-Italy Bilateral Symposium on the Coastal Zone and Continental Shelf Evolution Trend, October 5–8, 2010, Bologna, Italy (2010)
11. Costa, J.E.: Paleohydraulic reconstruction of flash-flood peaks from boulder deposits in the Colorado Front Range. Bull. Geol. Soc. Am. **94**, 986–1004 (1983)
12. Emmet, W.W.: A field calibration of sediment-trapping characteristics of the Helley-Smith bed-load sampler. USDI, GS Open File Report **79**, 79–411 (1979)
13. Grant, G.E., Schmidt, J.C., Lewis, S.L.: A geological framework for interpreting downstream effects of dams on rivers. In: O'Connor, J.E., Grant, G.E. (eds.) A Peculiar River: Geology, Geomorphology, and Hydrology of the Deschutes River, Oregon, pp. 203–219. AGU, Washington DC (2003)
14. Hammond, F.D.C., Heathershaw, A.D., Langhorne, D.N.: A comparison between Shields' threshold criterion and the movement of loosely packed gravel in a tidal channel. Sedimentology **31**, 51–62 (1984)
15. Herschy, R.W.: Streamflow Measurement. Elsevier, London (1985)

16. Hickin, E.J.: Rivers. http://www.sfu.ca/~hickin/RIVERS (2010), pp. 70–107
17. Hooke, J.M.: Human impacts on fluvial systems in the Mediterranean region. Geomorphology **79**, 311–335 (2006)
18. Inman, D.L., Jenkins, S.A.: Climate change and episodicity of sediment flux of small California rivers. J. Geol. **107**, 251–270 (1999)
19. Julien, P.Y.: Erosion and Sedimentation. Cambridge University Press, Cambridge (1995)
20. Kondolf, G.M.: Hungry water: effects of dams and gravel mining on river channels. Environ. Manag. **21**(4), 533–551 (1997)
21. Liebault, F., Piegay, H.: Assessment of channel changes due to long-term bedload supply decrease, Roubion River, France. Geomorphology **36**, 167–186 (2001)
22. Mennella, C.: Il clima d'Italia, vol. II, F.lli Conte Editori, Napoli, Italia (1972)
23. Parker, C., Clifford, N.J., Thorne, C.R.: Understanding the influence of slope on the threshold of coarse grain motion: revisiting critical stream power. Geomorphology **126**, 51–65 (2011)
24. Shields, A.: Application of similarity principles and turbulence research to bed-load movement. California Institute of Technology, Pasadena (Translated from German) (1936)

Part III
Theme: Glaciology

Mathematical Modeling of Rock Glacier Flow with Temperature Effects

Krishna Kannan, Daniela Mansutti, Kumbakonam R. Rajagopal, and Stefano Urbini

Abstract This paper stems from the interest in the numerical study of the evolution of Boulder Clay Glacier in Antarctica, whose morphological characteristics have required the revision of the basis for most of the recent mathematical models for glacier dynamics. Bearing in mind the need to minimize the complexity of the mathematical model, we have selected the constitutive equation of rock glacier ice recently presented by two of the authors. Here, this model is extended in order to include temperature effects. In addition to the effects of climate change, it is also necessary to take into consideration the non-negligible level of melting due to temperature changes induced by normal stresses arising from the interactions of ice and the rock fragments that are within the rock glacier. In fact, local phase transition that occurs leading to the release of water implies significant modifications of ice viscosity, the main intrinsic factor driving the flow.

In this paper we derive the model that describes the flow of rock glaciers that takes into consideration the effects of temperature and the normal stresses generated by the ice and rock fragments interactions.

Keywords Moraine · Rock fragments · Ice flow · Mixture · Normal stress differences

K. Kannan
Department of Mechanical Engineering, Indian Institute of Technology Madras, Chennai, India
e-mail: krishnakannan@iitm.ac.in

D. Mansutti
Istituto per le Applicazioni del Calcolo 'M. Picone', CNR, Rome, Italy
e-mail: d.mansutti@iac.cnr.it

K. R. Rajagopal (✉)
Department of Mechanical Engineering, Texas A&M University, College Station, TX, USA
e-mail: krajagopal@tamu.edu

S. Urbini
Istituto Nazionale di Geofisica e Vulcanologia, Rome, Italy
e-mail: stefano.urbini@ingv.it

© Springer Nature Switzerland AG 2020
P. Cannarsa et al. (eds.), *Mathematical Approach to Climate Change and Its Impacts*, Springer INdAM Series 38,
https://doi.org/10.1007/978-3-030-38669-6_4

149

1 Introduction

In the study of glacier evolution, in view of the high degree of complexity due to the numerous interacting processes, mathematical modelers typically seek a relatively simple constitutive equation that includes, to a reasonable approximation, those features of constitutive behavior which are important for the flow conditions of interest. The model proposed herein has been motivated by a real case study.

An Aircraft Runway at Boulder Clay Glacier We are interested in studying the evolution of the area of the Boulder Clay Glacier (BCG), laying on the flanks of the Northern Foothills in Victoria Land, facing Terra Nova Bay, in Antarctica, as in Fig. 1.

BCG is oriented parallel to the coast, elongated for 6 km from south to north, being about 1.5 km wide with average longitudinal elevation of about 50 m. Its thickness ranges between 100 m and 150 m and it results to be dry based. Actually, in the Northern Foothills, fluvial processes are relatively unimportant, the stream channels are extremely rare and an overall limited groundwater movement is observed. From this fact, we deduce that the 6–8 summer rainy weeks affect only bulk ice rheology; on the other hand annual external temperature averages to −30 °C, at most.

Within the rectangle in Fig. 1 it is apparent that BCG has three outlet glaciers which pull the main glacier branch by applying local moderate dynamical stress. The mingled white and grey zones are morainic deposits: adjacent to BCG on the east side, there is the debris-sized ice-cored Boulder Clay Moraine (BCM), which

Fig. 1 Satellite image of the area of Boulder Clay Glacier (BCG) and Boulder Clay Moraine (BCM) at Victoria Land, facing Terra Nova Bay in Antarctica (center image 74°43′ S, 164°01′ E)

occupies more than half of the total glacier's surface. Closer to the coastline, the bench of dark grey zones and wide icy spots degrading to the sea is partly made of deterioration moraines (of marine origin) and partly of granitic bedrock. The cliff margin is 30 m above see level (asl).

Across the second outlet glacier an aircraft runway (in Fig. 1 indicated as 'The Airstrip') is planned to serve the Italian Mario Zucchelli Station (north side) and other close polar stations. Our final task will be to estimate by numerical simulation either the impact of climate change on such an icy ground and the magnitude of possible deformations that may induce the onset of dangerous instabilities. In fact, available on-field measurements are providing not desiderable non-zero deformation values in that area [21].

Mathematical Models for Computational Glaciology Most existing literature in computational glaciology is based on the representation of ice as a newtonian power law fluid with the adoption of Glen's law [7] as a constitutive equation. In this case ice is described as a dense viscous fluid with viscosity coefficient, μ, proportional to a power law of the trace of the square of the strain-rate tensor, $\mathbf{A}_1$. Being $\mathbf{A}_1 = grad(\mathbf{v}) + (grad(\mathbf{v}))^T$ with $\mathbf{v}$, the ice Eulerian velocity field, Glen's law is based on the following generic expression for the viscosity coefficient

$$\mu = \mu_0[tr(\mathbf{A}_1^2)]^{\frac{m}{2}}, \tag{1}$$

with the factor μ_0 of phenomenological character and constant exponent m. This model, including the velocity gradient effect onto ice viscosity, is capable to represent the occurrence of secondary creep [12].

In Fig. 1 we observe that moraines concentrate at the end (first type) and aside (second type) of the ice flow path. The first type of deposits may have been pushed ahead along with the ice flow forming the *terminal moraines*, or, induced by the appearance of an obstacle (as in the case of stiff ice around the glacier margin), debris may have been up-warped from the glacier bedrock towards the surface leading to the formation of *inner moraines*, that may show up at surface as ridges, so-called *shear moraines* [2]. The second type of deposits are produced by lateral erosion of the bedrock, being kept and pushed aside along with flow of shearing ice. Each described mechanisms is influenced by the typical non-newtonian behavior occurring in rectilinear shearing flow, supported by non-zero normal stress differences, that qualitatively sums up to a thrust, carrying debris along, in each direction normal to the shear direction.

As of mathematical modelling, normal stress differences are both non identically zero for differential type fluid models of grade n, with $n > 1$. This is not true for the Navier-Stokes fluid model, that are of differential type with $n = 1$ [19], as well as the Glen's law which, consequently, cannot predict satisfactorily the whole process of moraine formation.

McTigue et al. [12] stressed that a good mathematical description of the creep of ice is provided by the second grade fluid model (SGFM) (or second-order fluid model, SOFM), the simplest constitutive equation capable of describing both primary and secondary creep, which also includes normal stress differences arising in shearing flow. For this model these authors have provided an estimate of the material's constitutive laws coefficients by comparison with biaxial creep experimental tests; then, via Schowalter's formula [19], they have also computed the maximum surface deflection for ice flow in a semi-circular channel as a function of the second normal stress difference. However the real impact of the second order terms of the model in natural glacier flows remained to be ascertained.

Along this line Man and Sun in [11] modified SOFM with a Glen-like functional expression of the coefficient (power law) of the first order term (MSOFM), in order to regain the strain rate effects as in the Glen's law. This resulted in less pronounced, more realistic normal stress difference effects than those estimated by McTigue et al. with the indication of the exception of extremely thick glaciers with steep slope; Man and Sun suggested also that lowering of temperature might greatly enhance normal stress effects in the creep of ice.

For the sake of completeness, we recall also the Elastic Modified Second Order Isotropic Material (EMSOIM) model, proposed by Reisen et al. [17], which is relevant to the description of ice response to an external load applied and then removed, as it is in the case of a forming and then draining supra-glacial lake. It is obtained by adding a Hooke-type elasticity term to the MSOFM with power law viscosity, characterized by vanishing bulk modulus with weakening elastic feedback on long time scales. EMSOIM model is also appropriate to model glacier flow with alternate extensional and compressional regimes, where respectively downhill and flat layered ice act alike a removed and applied load, or to capture the reversal of ice motion across a large subglacial lake [9] or in proximity of a subglacial high carrying capacity water flow [20].

Let us stress that above mentioned models are essentially for clean ice. In the case of rock or debris-laden glaciers (see sketch at Fig. 2), the impact of pressure and temperature must be also included in relation to the concentration of debris and particle size. Actually, here, within the flow of the glacier, ice melting temperature undergoes local changes due to (non-hydrostatic) pressure increase for the interaction of ice with included rock fragments, and subsequent local phase transition events lead to release of water influencing debris-ice mixture viscosity [5]. In this picture an important role is clearly played also by solute concentration (salinity) [1]. This behavior has been recently reviewed by Moore in [14] who reconsidered a broad range of field observations, theory and experimental work relevant to the mechanical interactions between ice and rock debris. It is worth mentioning the observations reported by Lawson and Elliott in [10] from experimental strength testing of real glacier ice sampled from Taylor Glacier in Antarctica for ice-cored debris and for clean ice. Lawson and Elliott found that, if debris is incorporated, transition from ductile to brittle regime (when crevasses may form) strongly depends on temperature, so that if temperature decreases, critical total strain also decreases, meanwhile compressive strength of rock glacier ice increases to its maximum as effect of regelation of previously formed water films, hardening around rocky particles. They observed also that, at low strain rates, as it is presumably in the present study case, the peak stress supported by ice in ductile

Fig. 2 Sketch of bedrock and
rock glacier

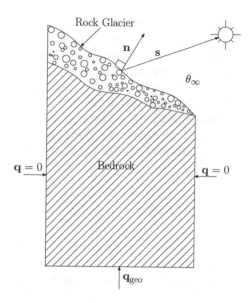

regime appears to increase with strain rate. What is relevant for the case study
that is considered is that at low temperatures debris-ice mixtures are usually more
resistant to deformation than their pure components; whereas, at temperatures close
to melting, the growth of unfrozen water films at ice-particle interfaces can lead
to pervasive weakening. The mechanism shaping ice-debris mixtures constitutive
behavior is summarized by Moore [14] as a "competition between the role of debris
in impeding ice creeping" (as for 'locking' in the case of granular material) "and
the mitigating effect of unfrozen water at debris-ice interface" driven by shear rate,
pressure, temperature, salinity, debris concentration and particle size.

Rock glaciers can be very unstable if a shear zone, adjacent to the bedrock,
establishes with effective viscosity that may get seven times smaller than for clean
ice: in this case they may start to move rigidly by fast shearing on the bottom with
possible devastating effects.

In 2013 two of the authors, Kannan and Rajagopal, proposed a generalization
of the Man and Sun's constitutive equation, MSOFM, by including, at mechanical
level, the presence of debris trapped in the ice as in a rock glacier. Similar to Mills
and Snabre [13], who modelled the flow of a suspension of hard spheres in a Newtonian viscous fluid, they enforced dissipation in the ice-debris mixture considering
clean ice interaction with debris particles and Coulombic friction between debris
particles within a shear flow at constant volume, either at *locking point* or away
from it. This led to an extension of the Mills and Snabre's relationships amongst
the shear and normal stresses of the suspension and the volume fraction of the
particles to the case of the icy mixture, obtaining a modified expression for the
viscosity function [8]. Furthermore, similar to the case of non-colloidal suspensions
faced by Morris and Boulay [15], the effects of debris particles volume fraction
on the other material coefficients are also recovered. Kannan and Rajagopal tested

their model on the motion of Murtel-Corvatsch rock glacier and pointed out that the inclusion of temperature effects would greatly improve the mathematical modeling of the process. In this paper we propose an extension of the Kannan and Rajagopal's model that includes the description of the thermal field, not only via the equation for the energy balance law but also through its impact on the constitutive equation of the ice and debris mixture.

The structure of this paper is the following: in Sect. 2, the balance equations relevant to the motion of a glacier are recalled; in Sect. 3, the extended constitutive assumptions are documented and motivated. Conclusions are drawn in Sect. 4.

2 Preliminaries

In this section we briefly record the balance equations for mass, linear and angular momentum, and energy that is relevant to the problem described by Fig. 2. Let us stress that rock and ice mixture is treated as a single constituent and the bedrock as a rigid sub-structure in comparison to the rock-glacier which, consisting of a mixture of ice and rock, essentially flows due to the external stimuli, albeit exceedingly slowly. Then, at the time scale of interest here, the bedrock meets identically mass and momentum conservation laws, and the related equations will be considered below only for the case of the rock glacier.

Balance of Mass Since the material is assumed to be incompressible, the consequence of the balance of mass is given by

$$\text{div } \mathbf{v} = 0, \tag{2}$$

where $\mathbf{v}$ is the velocity of the rock glacier (rock and ice mixture).

Balance of Linear Momentum Balance of linear momentum of the rock and ice mixture is given through

$$\text{div } \mathbf{T}^T + \rho_{\text{mix}}\mathbf{b} = \rho_{\text{mix}}\dot{\mathbf{v}} \tag{3}$$

where, $\mathbf{T}$ is the Cauchy stress, ρ_{mix} is the current density of the mixture and $\mathbf{b}$ is the specific body force.

Consequence of Balance of Angular Momentum In the absence of body couples, the consequence of balance of angular momentum is that the Cauchy stress is symmetric, i.e.

$$\mathbf{T}^T = \mathbf{T}. \tag{4}$$

Balance of Energy

Bedrock The balance of energy for a rigid body undergoing a thermal process in the absence of radiation, is given by

$$\rho_{br}\dot{\epsilon} + \text{div } \mathbf{q} = 0, \tag{5}$$

where, ρ_{br}, ϵ, $\mathbf{q}$ are the current density, specific internal energy and the heat flux vector, respectively. Constitutive assumption for the heat flux vector is given by Fourier's law of heat conduction as

$$\mathbf{q} = -k_{br} \text{ grad}(\theta), \tag{6}$$

where k_{br} is the thermal conductivity of the bedrock and θ its temperature.

Rock and Ice Mixture The balance of energy for the rock and ice mixture is given through

$$\rho_{\text{mix}}\dot{\epsilon} + \text{div } \mathbf{q} = \text{tr } (\mathbf{T} \cdot (\text{grad}\mathbf{v})^T) + \rho_{\text{mix}}\text{r} \tag{7}$$

where, ϵ is the current specific internal energy, r is the specific rate of supplied radiant heating and $\mathbf{q}$ is the current heat flux vector

$$\mathbf{q} = -k_{mix} \text{ grad}(\theta) \tag{8}$$

with k_{mix} and θ, the thermal conductivity of rock and ice mixture and its temperature respectively

3 Constitutive Assumptions

The above equations must be completed by constitutive relations for the mixture of rock and ice, as we account for the rock and sand grains trapped in the interstices of the rock glaciers. This is the subject of present section.

We approach, first, the thermodynamical quantities required in (7) and (8), as they also justify the introduction of new functions and parameters specifically needed when temperature effects are included.

From here on, we add self-explanatory subscripts related to the different materials (ice, rock and mixture of rock and ice).

3.1 Internal Energy for the Mixture

The specific internal energy for the rock and ice mixture in Eq. (7) is assumed of the form

$$\epsilon_{mix} = \left((1 - \phi)C^{ice} + \phi_{max}f\, C^{rock}\right)\theta \tag{9}$$

with

$$f = \frac{\bar{\phi}}{\bar{\phi} + e(1 - \bar{\phi})}, \tag{10}$$

where ϕ represents the volume fraction of the rock and sand grains trapped within the interstices of the rock glaciers, ϕ_{max} is the maximum volume fraction achievable in random-close-packing configuration, $\bar{\phi}$ is the relative volume fraction defined by

$$\bar{\phi} = \frac{\phi}{\phi_{max}}, \tag{11}$$

e is the parameter related to the extent of locking of the 'free-to-move' rock particles, C^{ice} and C^{rock} are the specific heat capacities of ice and rock, respectively, and θ is the absolute temperature.

Formula (9) through the quantity f in (10) calibrates the contribution of rock particles to specific internal energy by discriminating the effect of 'caged' particles, represented by $\bar{\phi}$, and the effect of 'free-to-move' rock particles, represented by $1 - \bar{\phi}$. This contribution increases proportionally up to ϕ_{max} either when the extent of locking, e, exerted by ice decreases (down to 0) and/or $\bar{\phi}$ increases (up to 1). The latter condition corresponds to either increased locking effect of ice (e.g. for thinning of ice inter-rock particle layer) and/or to increased number of 'caged' rock particles. In both cases at most the effect of the random-close-packing configuration can be reached.

The quantity f plays the role of the so-called equilibrium solid fraction in Mills and Snabre's theory of a concentrated suspension of non-Brownian hard spheres dispersed in a newtonian fluid [13]; actually, here, it represents the fraction of 'effectively locked' rock and ice suspension.

Let us notice that by setting e = 1, formula (9) identifies the simple rule for a two-phase mixture:

$$\epsilon_{mix} = \left((1 - \phi)C^{ice} + \phi\, C^{rock}\right)\theta, \tag{12}$$

where the contribution of each phase simply adds because the locking effect of ice upon 'free-to-move' rock particles is negligible.

Assuming that the specific heat capacities of the individual constituents are function only of temperature, the following relations are implicit:

$$C^{ice} = C_V^{ice} = C_T^{ice} \tag{13}$$

$$C^{rock} = C_V^{rock} = C_T^{rock}, \tag{14}$$

where, subscripts T and V denote the specific heat capacities with respect to constant stress and constant volume, respectively.

3.2 Heat Flux Associated with the Mixture

Similar to the effects on the internal energy described in Sect. 3.1, mixture thermal conductivity in Eq. (8) is also influenced by extent of the locking parameter, e, and is assumed of the form

$$k_{mix} = (1 - \phi)k^{ice} + \phi_{max} f k^{rock}. \tag{15}$$

3.3 Internal Heating of the Mixture

The internal heating of the mixture in Eq. (7) is shown (see Appendix) to be of the form:

$$\rho_{mix} r = \frac{I_p}{z}, \tag{16}$$

where I_p is the day-averaged potential direct solar radiation received by a surface element of the rock glacier and z is the thickness of the top ice layer most influenced by the solar radiation.

3.4 Cauchy Stress for the Mixture

Concerning the constitutive model for rock glacier motion, in the introductory review of glaciological models in Sect. 1, we had cited the work of Kannan and Rajagopal [8], that is extended in order to include the thermal effect.

Thus, the Cauchy stress is assumed to be of the form

$$\mathbf{T} = -p\mathbf{I} + \tilde{\mu}(\hat{p}, \theta)[a_0 + \frac{1}{2}\text{tr}(\mathbf{A}_1^2)]^{m/2}\mathbf{A}_1 + \tilde{\alpha}_1(\bar{\phi}, \theta)\mathbf{A}_2 + \tilde{\alpha}_2(\bar{\phi}, \theta)\mathbf{A}_1^2, \tag{17}$$

where p is the indeterminate part of the stress due to the incompressibility constraint, **I** is the identity tensor, $\mathbf{A_1}$ and $\mathbf{A_2}$ are the first two Rivlin-Ericksen tensors describing the symmetric part of the velocity gradient and its frame-indifferent time derivative [18], that is

$$\mathbf{A}_1 = \mathbf{L} + \mathbf{L}^T$$

$$\mathbf{A}_2 = \dot{\mathbf{A}}_1 + \mathbf{A}_1\mathbf{L} + \mathbf{L}^T\mathbf{A}_1$$

with **L**, the velocity gradient tensor and $\dot{\mathbf{A}}_1$, the material time derivative of tensor $\mathbf{A}_1$.

Furthermore:

$$\tilde{\mu}(\hat{p}, \theta) = \mu(\hat{p}) \exp\left[B\left(\frac{1}{\theta} - \frac{1}{\theta_0}\right)\right] \tag{18}$$

$$\tilde{\alpha}_i(\bar{\phi}, \theta) = \alpha_i(\bar{\phi}) \exp\left[B_i\left(\frac{1}{\theta} - \frac{1}{\theta_0}\right)\right], \quad i = 1, 2 \tag{19}$$

$$\theta_0 \text{ is the reference temperature} \tag{20}$$

$$\mu(\hat{p}) = \mu_0\left[(1-f)\left(1 + k_1\sqrt{\frac{\hat{p} - P_a}{P_a}}\right) + fk_2\frac{\hat{p} - P_a}{P_a}\right] \tag{21}$$

$$\alpha_i(\bar{\phi}) = \alpha_i^0\left[1 + k_{i+2}\frac{\bar{\phi}^2}{(1 - \bar{\phi})^2}\right], \quad i = 1, 2, \tag{22}$$

with $\hat{p} = -\frac{1}{3} tr(T)$ the negative of the mean normal stress.

Let us observe that the temperature dependence of the material functions $\tilde{\mu}, \tilde{\alpha}_1$ and $\tilde{\alpha}_2$, given in Eqs. (18) and (19), is assumed to be of Arrhenius form, in keeping with experimental observations [3]. When $\theta = \theta_0$, the iso-thermal case is recovered with those functions reduced to the expressions for $\mu(\hat{p})$ and $\alpha_i(\bar{\phi})$ ($i = 1, 2$) in (21) and (22), respectively. The latter ones are justified in the work by Kannan and Rajagopal [8].

We require that the Cauchy stress is thermodynamically consistent in the sense that the model meets the second law of thermodynamics and the Helmholtz potential is a minimum in equilibrium (see Dunn and Rajagopal [4]), which leads to:

$$\mu \geq 0, \alpha_1 \geq 0 \text{ and } \alpha_1 + \alpha_2 = 0. \tag{23}$$

These conditions led Kannan and Rajagopal to fix the quantities in (22) such that $\alpha_1(\bar{\phi}) + \alpha_2(\bar{\phi}) = 0$, that is met by $k_3 = k_4$ and $\alpha_1^0 + \alpha_2^0 = 0$. Here, by including $B_1 = B_2$, the required identity

$$\tilde{\alpha}_1(\bar{\phi}, \theta) + \tilde{\alpha}_2(\bar{\phi}, \theta) = 0 \tag{24}$$

is met for the present case too.

It is worth noticing that the constitutive equation (17) includes the MSOFM formulation by Man and Sun [11], which is obtained for $a_0 = 0$ and constant value for $\tilde{\mu}(\hat{p}, \theta)$, $\tilde{\alpha}_i(\bar{\phi}, \theta)$ $(i = 1, 2)$, and also Glen's power law model resulting with $a_0 = \tilde{\alpha}_i(\bar{\phi}, \theta) = 0$ $(i = 1, 2)$. Actually, the assignment of the value of the exponent m is done according to Glen's law experimentation results [3] that lead to $m = -2/3$ for the class of glacier flow problems considered here.

For glaciers characterized by small elevation gradient, as it is the case of BC glacier, elastic effects are not expected, then, the Hooke's type elasticity term of EMSOIM formulation is not necessary.

4 Concluding Observations

A number of parameters have to be still fixed in order to fit the natural set-up of the glacier dynamics problem.

e, ϕ and ϕ_{max} pertain to the composition of the mixture of ice and debris (ice-rock particles binding effect, volume fraction).

$\mu_0, \alpha_1^0, \alpha_2^0, k_1$ and k_2 as well as k_3 and k_4 are material constants: μ_0, α_1^0 and α_2^0 are viscosity and normal stress moduli, respectively, of pure ice at atmospheric pressure and constant temperature θ_0. The k_i are a measure of the impact of the mixture composition on viscosity μ and material moduli α_1, α_2. As suggested by Kannan and Rajagopal [8], there is no apparent reason against the choice $k_1 = k_2 = k_3 = k_4$.

In general, in absence of laboratory estimates, for choosing a value for each parameter, a 'trial and error' procedure is suggested in order to obtain the best fit of consolidated experimental measures of observable quantities.

Finally, a paper on the numerical experimentation of the presented model is in preparation. In particular we are considering the Murtel Corvatsch glacier flow versus in-depth borehole deformation measurements in order to compare with the accuracy obtained by adopting Kannan and Rajagopal's isothermal model. Furthermore, a simplified geometry for the Boulder Clay Glacier and Moraine will be considered for approaching the problem discussed in the Introduction.

Acknowledgements The presented Boulder Clay Glacier problem is part of the research plan of the project ENIGMA: authors acknowledge Piano Nazionale Ricerca Antartide (PNRA) for partial financial support of this topic (project PNRA16-00121). KR acknowledges also the financial support received by the Italian Ministry of Instruction, University and Research by the funding MATHTECH granted to the Istituto per le Applicazioni del Calcolo 'M. Picone' (CNR).

Appendix

In general, heating of a rock glacier due to solar radiation is assumed to be a surface phenomenon. Representing the topography of a surface through a digital terrain model, Funk and Hoelzl [6] calculated a day-averaged potential direct solar radiation, I_p, received by a surface element through the relation

$$I_p = \int_{t_{sr}}^{t_{st}} I_0 \, \mathbf{s} \cdot \mathbf{n} \, dt \tag{25}$$

where the flux-like quantity at a given point on the surface, I_p, depends on the intensity of solar radiation, I_0, received at the point and the two directional vectors, $\mathbf{n}$ and $\mathbf{s}$, being the unit normal vector to the surface at the point and the unit directional vector oriented towards the sun from the given point, respectively. It is noted that I_p is time-averaged for a day, i.e. between the time of sunrise, t_{sr}, and the time of sunset, t_{st}. Such time-averaging for a day is assumed to be acceptable in view of the fact that time-integration involved in applicative numerical simulations relates to such large sized systems (in length and in time scale).

Temperature profiles of bored holes suggest that only a small layer of thickness below the exposed surface of rock glacier falls in the warmer regime in comparison to the deeper regions [16, 22]. If one assumes that such small layer is the most influenced region by the solar radiation received at the surface, then the heating due to solar energy, although a surface phenomenon, can be considered to be volumetric one, i.e. an equivalent expression for body heating can be determined from the flux equation; however the body under consideration will be restricted to a small layer of thickness, say up to a depth z, below the surface. By integrating I_p in Eq. (25) over the entire boundary $\partial\Omega$, one obtains the total flux and by converting this surface integral to a volumetric one, whose domain is given by $\Omega = \partial\Omega \times [0, z]$, the following relations are derived:

$$\int_{\partial\Omega} \underbrace{\left(\int_{t_{sr}}^{t_{st}} I_0 \, \mathbf{s} \cdot \mathbf{n} \, dt \right)}_{I_p} da = \int_{\Omega} \left(\int_{t_{sr}}^{t_{st}} \mathrm{div}(I_0 \, \mathbf{s}) dt \right) dv =$$

$$= \int_0^z \int_{\partial\Omega} \left(\int_{t_{sr}}^{t_{st}} \mathrm{div}(I_0 \, \mathbf{s}) dt \right) da \, dh = z \int_{\partial\Omega} \underbrace{\left(\int_{t_{sr}}^{t_{st}} \mathrm{div}(I_0 \, \mathbf{s}) dt \right)}_{\rho_{mix} \mathbf{r}} da.$$

By matching the integrands of the first and the last integral, after dividing by the thickness, z, of the ice layer most influenced by the solar radiation, the identity (16) is deduced

$$\frac{I_p}{z} = \rho_{mix} \, \mathbf{r}.$$

References

1. Chen, C.T., Millero, F.J.: Precise thermodynamic properties for natural waters covering only the limnological range. Limnol. Oceanogr. **31**(3), 657–662 (1986)
2. Chinn, T.J.: Polar glacier margin and debris features. Mem. Soc. Geol. It. **46**, ff. 16, 25–44 (1991)
3. Cuffey, K.M., Paterson, W.S.B.: The Physics of Glaciers. BH Elsevier, Croydon (2010)
4. Dunn, J., Rajagopal, K.R.: Fluids of differential type: critical review and thermodynamic analysis. Int. J. Eng. Sci. **33**(5), 689–729 (1995)
5. Duval, P.: The role of the water content on the creep rate of polycrystalline ice. IAHS Publ. **118**, 29–33 (1997)
6. Funk, M., Hoelzle, M.: A model of potential direct solar radiation for investigating occurrences of mountain permafrost. Permafr. Periglac. Process. **3**(2), 139–142 (1992)
7. Glen, J.W.: The creep of polycrystalline ice. Proc. R. Soc. Lond. A Math. Phys. Eng. Sci. **228**(1175), 519–539 (1955)
8. Kannan, K., Rajagopal, K.R.: A model for the flow of rock glaciers. Int. J. Non-linear Mech. **48**, 59–64 (2013)
9. Kwok, R., Siegert, M., Carsey, F.: Ice motion over Lake Vostok, Antarctica: constrains in inferences regarding the accreted ice. J. Glaciol. **46**, 689–694 (2000)
10. Lawson, W., Elliott, C.: Strain rate effects on the strength of debris-laden glacier ice. N. Z. J. Geol. Geophys. **46**, 323–330 (2010)
11. Man, C.-S., Sun, Q.-S.: On the significance of normal stress effects in the flow of glaciers. J. Glaciol. **33**(115), 268–273 (1987)
12. McTigue, D.F., Passman, S.L., Jones, S.J.: Normal stress effects in the creep of ice. J. Glaciol. **31**(108), 120–126 (1985)
13. Mills, P., Snabre, P.: Apparent viscosity and particle pressure of a concentrated suspension of non-Brownian hard spheres near the jamming transition. Eur. Phys. J. E **30**(3), 309–316 (2009)
14. Moore, P.L.: Deformation of debris-ice mixtures. Rev. Geophys. **52**, 435–467 (2014)
15. Morris, J.F., Boulay, F.: Curvilinear flows of noncolloidal suspensions: the role of normal stresses. J. Rheol. **43**, 1213–1237 (1999)
16. Myhra, K.S., Westermann, S., Etzelmüller, B.: Modelled distribution and temporal evolution of permafrost in steep rock walls along a latitudinal transect in Norway by CryoGrid 2D. Permafr. Periglac. Process. **28**(1), 172–182 (2017)
17. Riesen, P., Hutter, K., Funk, M.: A viscoelastic Rivlin-Ericksen material model applicable to glacier ice. Nonlinear Process. Geophys. **17**, 673–684 (2010)
18. Rivlin, R.S., Ericksen, J.L.: Stress-deformation relations for isotropic materials. J. Rational Mech. Anal. **4**, 323–425 (1955)
19. Schowalter, W.R.: Mechanics of Non-Newtonian Fluids. Pergamon Press, London (1978)
20. Sugiyama, S., Gudmundsson, G.H.: Short-term variations in glacier flow controlled by subglacial water pressure at Lauteraargletscher, Bernese Alps, Switzerland. J. Glaciol. **50**(170), 353–362 (2004)
21. Urbini, S., Bianchi-Fasani, G., Mazzanti, P., Rocca, A., Vittuari, L., Zanutta, A., Alena Girelli, V., Serafini, M., Zirizzotti, A., Frezzotti, M.: Multi-temporal investigation of the Boulder Clay Glacier and Northern Foothills (Victoria Land, Antarctica) by integrated surveying techniques. Remote Sens. **11**, 1501 (2019). https://doi.org/10.3390/rs11121501
22. Zhou, X., Buchli, T., Kinzelbach, W., Stauffer, F., Springman, S.M.: Analysis of thermal behaviour in the active layer of degrading mountain permafrost. Permafr. Periglac. Process. **26**(1), 39–56 (2015)

A Model to Describe the Response of Arctic Sea Ice

Reza Malek-Madani and Kumbakonam R. Rajagopal

Abstract In this paper we develop a model for the flow of Arctic sea ice within the context of the theory of interacting continua that takes into account the change of phase between the two constituents, ice and water. After documenting the general balance laws for mass, linear and angular momentum, energy, the second law of thermodynamics and the volume additivity constraint, we discuss the specific constitutive relations that are to be used for the various quantities that appear in the balance laws. Ice is modeled as a non-Newtonian fluid that is a generalization of the usual model due to Glen to take into account the ability of ice to develop normal stress differences in simple shear flow, while water is modeled as a Navier–Stokes fluid. Constitutive relations are discussed for the change of phase, the interaction forces such as the drag, etc. In order to make the problem amenable to analysis, we simplify the governing equations, keeping the quintessential features of the problem of interest in mind.

Keywords Arctic sea ice · Ice-water mixture · Theory of mixtures · Second law of thermodynamics · Glen's law

1 Introduction

The behavior of sea ice in the Arctic is considered one of the key indicators of climate variability and its decline in the past few decades is presenting social, political, economic and ecological challenges. Predicting the sea ice extent is currently one of the important open problems in geophysical fluid dynamics, and because

R. Malek-Madani (✉)
Department of Mathematics, US Naval Academy, Annapolis, MD, USA
e-mail: rmm@usna.edu

K. R. Rajagopal
Department of Mechanical Engineering, Texas A&M University, College Station, TX, USA
e-mail: krajagopal@tamu.edu

© Springer Nature Switzerland AG 2020
P. Cannarsa et al. (eds.), *Mathematical Approach to Climate Change and Its Impacts*, Springer INdAM Series 38,
https://doi.org/10.1007/978-3-030-38669-6_5

163

of its unique geographical setting, there is now an abundance of data available for modelers to investigate and improve the analytical and computational models of sea ice formation and its deformation. The review paper by Perovich and Richter-Menge [27] outlines the observed reduction of sea ice due to thermodynamic and dynamic processes, while Bushuk et al. [5] study the reemergence of sea ice based on recent data analysis algorithms that are capable of extracting detailed information about correlation between the spring and fall anomalies in sea ice extents. The above-mentioned papers are just two among many studies in the past decade that intend to quantify mathematically what sea ice is and how it interacts with its ocean and atmospheric boundaries.

In this paper, we shall develop a simplified reduced model for describing the motion of Arctic sea ice based on the theory of mixtures. The theory allows one to take into consideration the motion of complex mixtures wherein the constituents of the mixture can chemically react with one another. While the roots of the theory can be traced back to the seminal works of Fick [9] and Darcy [7], a fully three dimensional treatment of the basic balance laws for the various constituents of the mixture, within the context of continuum mechanics, was first introduced by Truesdell [38, 39]. Detailed discussion of the basic ideas of the theory can be found in [41] and in the review articles by Bowen [4], Atkin and Craine [1, 2], Bedford and Drumheller [3] and the books by Samohyl [36] and Rajagopal and Tao [33] and the various appendices in the book by Truesdell [40]. For a hierarchy of approximations of mixture theory that leads to models due to Darcy, Forchheimer, Brinkman, Biot, etc., see [29].

Mixture theory has been used to study problems concerning ice sheets (Hutter [14], Hutter et al. [15]), ice streams in ice sheets (see Marshall and Clarke [20, 21] and the references therein), and snowpacks (see Kelly et al. [17], Morland et al. [25], Morris [26] and the references therein). After documenting the balance laws for mass, linear and angular momentum, energy for each constituent, and the second law of thermodynamics for the mixture, we shall consider a very simple mixture model with just two constituents (which also happens to be two phases of a material rather than two distinct materials), one for the ice and the other for water, allowing for phase change to take place between these two constituents.

The basic assumption of mixture theory is that the various constituents of the mixture co-occupy a point in space, in a homogenized sense.[1] Of course, at any point in the region belonging to the mixture there is only one constituent present, but each constituent is homogenized, that is, assumed to be spread over the region occupied by the mixture (any arbitrary neighborhood around any point occupied by the mixture has sufficiently large amount of each constituent of the mixture), and each of these constituents have their own motion, which allows for a "diffusion

---

[1]Here, homogenization does not refer to any specific mathematical process of mathematical homogenization procedure but refers to the sense in which we think of a body as a single continuum though clearly at the sub-atomic level the body is really not a three dimensional continuum. Each constituent of the mixture is assumed to be present at each point in the region occupied by the mixture, we do not have regions where we have only one constituent and not the others.

velocity" (relative velocity) between the constituents. Kinematical quantities and a partial stress can be defined for each constituent of the mixture just as one does within the context of a single continuum. One can also define average quantities associated with the mixture and this can be carried out by weighting in a variety of ways. Mixture theory also allows one to take into account the chemical reactions that could take place amongst the constituents that can produce one or more of the constituents at the expense of others. With regard to the problem that interests us, such a theory can accommodate the phase change that can take place between the ice and water. We document the balance laws for mass, linear and angular momentum, energy and the second law of thermodynamics in the form of the Clausius–Duhem inequality.

There are fundamental questions concerning the application of the second law of thermodynamics even when considering a single constituent material: in what form should the second law be enforced, should one only enforce it as a global inequality holding in the whole body or could one enforce it in the local form, in a particular form to just closed systems, and many other such questions. When it comes to mixtures there is an additional consideration, namely whether the second law ought to be enforced on the mixture as a whole or whether the stronger assumption that the second law be met by each of the constituents.

Mixture theory is not without its attendant disadvantages. One of the thorniest problem within the context of mixtures is the assignment of boundary conditions (see the detailed discussions in Shi et al. [37], Rajagopal et al. [34]). For some problems, one only knows the total traction/displacement/velocity on the boundary of the mixture. How this is to be split up amongst the different constituents is far from clear. While in some specific problems, splitting of the boundary condition based on different physical considerations leads to very similar qualitative results (see Prasad and Rajagopal [28]), this might not be generally true. Here, we adopt a very simple approach to assigning the boundary conditions for the constituents.

It is worth noting at this point that the model we will develop in this paper is substantially different from the one developed in the important paper by Gray and Morland [11]. Specifically, the model that Gray and Morland used for ice is merely a power-law model and does not capture the response of ice in an appropriate manner. There is considerable evidence that ice exhibits normal stress differences in simple shear flow, which the power-law model is incapable of capturing. The model for ice presented here can explain normal stress differences in a simple shear flow. Also, Gray and Morland do not consider all the interaction terms that might arise in a mixture. More importantly, the aforementioned paper assumes from the very beginning a two-dimensional approach to the problem, in addition to several additional constraints such as an integrated constitutive relation for the extra stress of the ice-water mixture. Our aim is to develop a fully general set of governing equations within the context of mixture theory which could then be applied to ice-water mixtures in a variety of situations.

A second important and influential paper on sea ice dynamics is by Hunke and Dukowicz [13] where the model is motivated by the desire to obtain a system of partial differential equations that allows for fast and efficient numerical computation. Consequently a number of simplifying assumptions are introduced that are not based on the physics of the underlying problem, but do lead to a two-dimensional model that has lent itself to successful computations. Our goal here is to develop a fully general model that adheres to the known physical character of sea ice. Developing fast and efficient numerical methods to solve the new system of PDEs that captures the physics of the problem under consideration will be the goal of a future study.

2 The Basic Equations of Mixture Theory

We shall provide a brief review of the basic equations of mixture theory. While we shall be interested only in a mixture of ice and water, we shall discuss the general theory within the context of several interacting continua that constitute the mixture. Let us consider a mixture of N constituents. The basic tenet of mixture theory is that the N constituents can be homogenized and assumed to co-occupy the domain occupied by the mixture. This allows for us to model the motion of the various constituents with respect to one another and in general the constituents can interact chemically to form new constituents. Of course, the formation of any constituent is at the expense of the other constituents as the balance of mass as a whole has to be met.

Now, at each point $\mathbf{x}$ belonging to the mixture at time t, which is occupied by a particle belonging to each of the homogenized constituents, we assign a density associated with each of the constituents. Let ρ^α denote the density of the αth constituent. In addition to assigning properties associated with each constituent, we can also define a quantity associated with that particular property for the mixture as a whole. As this definition is not unique, one has to be careful in interpreting the physical implications of these mixture quantities. The density ρ of the mixture can be defined through

$$\rho = \sum_{\alpha=1}^{N} \rho^\alpha. \tag{1}$$

Further, let $\mathbf{X}^\alpha$, $\alpha = 1, 2, \ldots, N$, denote a typical point belonging to each constituent in the reference state, which at the time t coexist at the point $\mathbf{x}$ that belongs to the mixture. The Greek alphabet superscript stands for the constituent and should not be confused with the tensorial component. The motion of the body is defined through

$$\mathbf{x} = \chi^\alpha(\mathbf{X}^\alpha, t), \quad \alpha = 1, 2, \ldots, N \tag{2}$$

or in component form

$$x_i = \chi_i^\alpha(\mathbf{X}^\alpha, t), \quad i = 1, 2, \ldots, n. \tag{3}$$

Unless explicitly stated, the usual Einstein summation convention will not apply in what follows.

We define the velocity of the αth constituent as

$$\mathbf{v}^\alpha = \frac{\partial \mathbf{x}^\alpha}{\partial t}, \tag{4}$$

or in indicial notation

$$v_i^\alpha = \frac{\partial x_i^\alpha}{\partial t}. \tag{5}$$

The mean velocity of the mixture is defined through

$$\mathbf{v} = \frac{1}{\rho} \sum_{\alpha=1}^{N} \rho^\alpha \mathbf{v}^\alpha. \tag{6}$$

The difference between the average velocity of the mixture and the velocity of a component is a measure of how the component is diffusing through the mixture as a whole and we define the diffusion velocity $\mathbf{v}_d^\alpha$ of the αth constituent through

$$\mathbf{v}_d^\alpha = \mathbf{v} - \mathbf{v}^\alpha. \tag{7}$$

With regard to mixtures various derivatives can be used depending on whether we are following a constituent or the mixture as a whole. We shall use the following notation:

$$\phi'^\alpha = \frac{\partial}{\partial t} \phi^\alpha(\chi^\alpha(\mathbf{X}^\alpha, t), t) = \frac{\partial \phi^\alpha}{\partial t} + [\frac{\partial}{\partial \mathbf{x}} \phi^\alpha(\mathbf{x}, t)] \mathbf{v}^\alpha(\mathbf{x}, t), \tag{8}$$

$$\frac{d\phi}{dt} = \frac{\partial \phi}{\partial t} + [\frac{\partial}{\partial \mathbf{x}} \phi(\mathbf{x}, t)] \mathbf{v}(\mathbf{x}, t). \tag{9}$$

We can define the acceleration $\mathbf{a}^\alpha$, velocity gradient $\mathbf{L}^\alpha$, and the deformation gradient $\mathbf{F}^\alpha$ for the αth constituent through

$$\mathbf{a}^\alpha = \frac{\partial^2 \mathbf{x}^\alpha}{\partial t^2}, \quad \mathbf{F}^\alpha = \frac{\partial \chi^\alpha}{\partial \mathbf{X}^\alpha}, \quad \mathbf{L}^\alpha = \frac{\partial \mathbf{v}^\alpha}{\partial \mathbf{x}}, \tag{10}$$

or in indicial notation

$$a_i^\alpha = \frac{\partial^2 x_i^\alpha}{\partial t^2}, \quad F_{ij}^\alpha = \frac{\partial x_i^\alpha}{\partial X_i^\alpha}, \quad L_{ij} = \frac{\partial v_i^\alpha}{\partial x_j^\alpha}. \tag{11}$$

The symmetric and skew parts of the velocity gradient $\mathbf{D}^\alpha$ and $\mathbf{W}^\alpha$ are defined through

$$\mathbf{D}^\alpha = \frac{1}{2}\left[\mathbf{L}^\alpha + (\mathbf{L}^\alpha)^T\right], \quad \mathbf{W}^\alpha = \frac{1}{2}\left[\mathbf{L}^\alpha - (\mathbf{L}^\alpha)^T\right], \tag{12}$$

or in indicial notation

$$D_{ij}^\alpha = \frac{1}{2}\left[L_{ij}^\alpha + L_{ji}^\alpha\right], \quad W_{ij}^\alpha = \frac{1}{2}\left[L_{ij}^\alpha - L_{ji}^\alpha\right]. \tag{13}$$

A partial traction $\mathbf{t}^\alpha$ is associated with each constituent and the corresponding partial stress associated with the partial tractions are denoted by $\mathbf{T}^\alpha$ and

$$\mathbf{t}^\alpha = (\mathbf{T}^\alpha)^T \mathbf{n}, \quad t_i^\alpha = T_{ji}^\alpha n_j, \tag{14}$$

where $\mathbf{n}$ is the unit outward normal to the surface on which the traction $\mathbf{t}$ acts. The total traction in the mixture has been defined in more than one way, one so that the form for the mixture traction would lead to a balance equation for the mixture to resemble that for a single constituent (see Truesdell and Toupin [41]), and the other as just the sum of the individual stresses associated with the constituents (see Green and Naghdi [12]). This difference in the definition of the total stress for the mixture is merely a matter of book-keeping, and the two definitions of total stress are consistent with one another. As we shall use the balance laws for the individual constituents, we shall not concern ourselves with this issue.

We shall not record the general integral form of the balance laws and then derive the local form since such treatments can be found in the review articles and books concerning mixture theory, and with regard to ice-water mixtures in the papers by Morland and co-authors [23–25], and Marshall and Clarke [20, 21], cited in the introduction.

3 Balance of Mass

The balance of mass for the αth constituent in local form, is given by

$$\frac{\partial \rho^\alpha}{\partial t} + \operatorname{div}(\rho^\alpha \mathbf{v}^\alpha) = m^\alpha \quad (\text{ no sum on } \alpha), \tag{15}$$

where m^α is the production of mass of the αth constituent due to the interaction amongst the constituents. On summing the above Eq. (15) over all constituents we obtain

$$\frac{\partial \rho}{\partial t} + \sum_{i=1}^{N} \frac{\partial(\rho \mathbf{v})_i}{\partial x_i} = 0. \tag{16}$$

We have used the fact that

$$\sum m_\alpha = 0, \tag{17}$$

as the net mass production of all the constituents of the mixture has to be zero, that is, if the mass of certain constituents increase, the mass of the others have to decrease by the exact same amount.

We could also express the balance of mass in the form:

$$\frac{d}{dt}(\rho^\alpha \det \mathbf{F}^\alpha) = m^\alpha \det \mathbf{F}^\alpha, \tag{18}$$

where $\frac{d}{dt}$ is the material time derivative.

4 Balance of Linear Momentum

The balance of linear momentum for the αth constituent is given by

$$\operatorname{div}(\mathbf{T}^\alpha)^T + \rho^\alpha \mathbf{b}^\alpha + \mathbf{m}^\alpha + m^\alpha \mathbf{v}^\alpha = \frac{\partial}{\partial t}(\rho^\alpha \mathbf{v}^\alpha) + \operatorname{div}(\rho^\alpha \mathbf{v}^\alpha \otimes \mathbf{v}^\alpha), \tag{19}$$

or in indicial form

$$\frac{\partial T_{ji}^\alpha}{\partial x_j} + \rho^\alpha b_i^\alpha + m_i^\alpha + m^\alpha v_i^\alpha = \frac{\partial(\rho^\alpha v_i^\alpha)}{\partial t} + \frac{\partial}{\partial x_i}(\rho^\alpha v_i^\alpha v_j^\alpha). \tag{20}$$

It is important to note that the scalar m^α is the production of mass while the vector $\mathbf{m}^\alpha$ is the supply of momentum due to the interaction between the constituents. It is also important to recognize that the term $\mathbf{m}^\alpha$ is a consequence of the momentum transfer due to the presence of the other constituents while $m^\alpha \mathbf{v}$ is the momentum supply due to the mass production of the αth constituent. Morland et al. [25] consider an interaction force that just consists of the drag force and a force associated with the phase change of the other constituents that is linear in the difference in the velocities between the constituents. It is not clear that this ought to be the form used or such a form brings in any new physical aspect into consideration as the drag also is directly proportional to the difference in the velocity between the constituents. We shall not adopt such an approach but choose to work with the general interaction force $\mathbf{m}^\alpha$ and the additional momentum due to phase change associated with the αth constituent $m^\alpha \mathbf{v}^\alpha$, which allows us to make changes to the constitutive assumption for $\mathbf{m}^\alpha$ on the basis of the specific issues on hand. We shall discuss the constitutive assumption for the interaction term $\mathbf{m}^\alpha$ in more detail later.

The total stress $\mathbf{T}$ associated with the mixture can be defined in a variety of ways. Here, we define the stress in the mixture through

$$\mathbf{T} = \sum_{\alpha=1}^{N} \mathbf{T}^{\alpha}, \quad T_{ij} = \sum_{\alpha=1}^{N} T_{ij}^{\alpha}. \tag{21}$$

On defining

$$\hat{\mathbf{T}} = \mathbf{T} - \sum_{\alpha=1}^{N} (\rho^{\alpha} \mathbf{u}^{\alpha} \otimes \mathbf{u}^{\alpha}), \quad \hat{T}_{ij} = T_{ij} - \sum_{\alpha=1}^{N} \rho^{\alpha} u_i^{\alpha} u_j^{\alpha} \tag{22}$$

and summing (19) over α, one obtains an equation (not explicitly documented here) that has the same structure as that for a single constituent, provided

$$\sum_{\alpha=1}^{N} (\mathbf{m}^{\alpha} + m^{\alpha} \mathbf{v}^{\alpha}) = \mathbf{0}. \tag{23}$$

5 Balance of Angular Momentum

The balance of angular momentum reduces to

$$\mathbf{M}^{\alpha} = (\mathbf{T}^{\alpha})^{T} - \mathbf{T}^{\alpha}, \quad M_{ij}^{\alpha} = T_{ji}^{\alpha} - T_{ij}^{\alpha}, \tag{24}$$

where $\mathbf{M}^{\alpha}$ denotes the angular momentum supply to the αth constituent. Thus, in general the partial stress tensor is not symmetric. In our study, we shall assume a symmetric partial stress tensor for the constituents and hence there is no angular momentum supply to the individual constituents.

6 Balance of Energy

In mixtures, one can associate a different temperature with regard to each constituent. However, as we are interested in a mixture whose constituents are undergoing slow flow, we can associate just one temperature with all the constituents. This simplifies the energy balance considerably. The balance of energy for the αth constituent is

$$\frac{\partial}{\partial t} \left[\rho^{\alpha} (\epsilon^{\alpha} + \frac{1}{2} \mathbf{v}^{\alpha} \cdot \mathbf{v}^{\alpha}) \right] + \text{div} \left[\rho^{\alpha} (\epsilon^{\alpha} + \frac{1}{2} \mathbf{v}^{\alpha} \cdot \mathbf{v}^{\alpha}) \mathbf{v}^{\alpha} \right] = \text{div} \left[\mathbf{T}^{\alpha} \mathbf{v}^{\alpha} - \mathbf{q}^{\alpha} \right] +$$

$$\rho^{\alpha} r^{\alpha} + \rho^{\alpha} \mathbf{v}^{\alpha} \cdot \mathbf{b}^{\alpha} + \epsilon_S^{\alpha} + m^{\alpha} (\epsilon^{\alpha} + \frac{1}{2} \mathbf{v}^{\alpha} \cdot \mathbf{v}^{\alpha}) + \mathbf{m}^{\alpha} \cdot \mathbf{v}^{\alpha}, \tag{25}$$

where ϵ^α is the internal energy of the αth constituent, ϵ_S^α is the energy supply to the αth constituent, $\mathbf{q}^\alpha$ is the heat conduction associated with α^{th} constituent and r^α is the radiant heat supply to the αth constituent.

7 Second Law of Thermodynamics

Even when attention is focused on enforcing the second law of thermodynamics in its local form, there are two different points of view in mixture theory. The first requires that the second law hold for each constituent while the other requires that it holds for the mixture as a whole. Even after one makes a decision with regard to how we decide to enforce the second law, it is important to recognize that there is not yet an agreement as to what form the second law should take, whether one should interpret it as the Planck inequality, Kelvin–Planck inequality, the Clausius inequality, the Clausius–Duhem inequality (see Coleman and Noll [6]), the approach adopted by Caratheodory, or the more recent choice wherein an constitutive assumption is made for the structure of the rate of entropy production which is required to be non-negative (see the approaches of Ziegler [42–44], and those of Rajagopal and Srinivasa [30–32]; the approaches are quite different though at first glance can be mistaken to be similar, see the extended discussion in Rajagopal and Srinivasa [31] for the differences between the two approaches), or a whole host of other approaches. We shall not get into a discussion of these issues here but use the simple approach that the second law is to be satisfied for the mixture as a whole and not the individual constituents, and furthermore enforce the second law in the form

$$\frac{\partial}{\partial t}(\sum_{\alpha=1}^{N}\rho^\alpha\eta^\alpha) + \text{div}(\sum_{\alpha=1}^{N}\rho^\alpha\eta^\alpha\mathbf{v}) + \text{div}(\sum_{\alpha=1}^{N}\frac{\mathbf{q}^\alpha + \rho^\alpha\eta^\alpha\theta^\alpha\mathbf{v}}{\theta^\alpha}) - \sum_{\alpha=1}^{N}\frac{\rho^\alpha r^\alpha}{\theta^\alpha} \geq 0,$$

(26)

where θ^α, $\mathbf{q}^\alpha$, η^α, and r^α are the absolute temperature, heat flux vector, specific entropy, and the specific radiant heating associated with the αth constituent.

A detailed discussion of the rationale for using the second law for the mixture as a whole, and the derivation and explanation of the terms that appear in the above equation can be found in Rajagopal and Tao [33].

8 Volume Additivity Constraint

In addition to the above balance laws and the entropy production inequality, just as in the case of a single constituent continuum we might have to enforce constraints such as incompressibility or inextensibility. A constraint that is often used in the theory of mixtures is that of the additivity of the volumes of the constituents (see Mills [22]). This is different from the requirement that each of the constituents is individually incompressible, it is a much weaker assumption and gives rise to just one Lagrange multiplier. Also, from a physical standpoint, allowing each constituent to be incompressible presents a fundamental problem as each constituent also occupies, in a homogenized sense, the whole space occupied by the mixture. This would mean that the mapping that takes the reference configuration for the pure constituent to that of the mixture cannot be iso-choric, or put physically one cannot spread a pure incompressible constituent from its reference state to the current state of the mixture which is the sum of the volume of all the constituents. As we shall just consider two constituents later, namely water and ice, we shall document the volume additivity constraint in the case of these two constituents:

$$\frac{\rho^I}{\rho_R^I}(1 - \beta) + \frac{\rho^W}{\rho_R^W} = 1, \tag{27}$$

where β is the porosity of ice, the superscripts I and W denote ice and water, and the subscript R denotes that the quantity in question refers to its value in the pure constituent configuration. The above can be described in the equivalent Eulerian form:

$$\frac{\rho^I}{\rho_R^I}\operatorname{div}\mathbf{v}^I + \frac{1}{1-\beta}\frac{\rho^W}{\rho_R^W}\operatorname{div}\mathbf{v}^W + \nabla(\frac{\rho^I}{\rho_R^I}) \cdot (\mathbf{v}^I - \mathbf{v}^W) = 0. \tag{28}$$

9 Issues Concerning Boundary Conditions for Mixtures

In order to solve the appropriate initial-boundary value problem, it is necessary to be able to prescribe appropriate boundary conditions. This is no easy task due to the assumption of co-occupancy. While one knows the total traction on the mixture, one does not know how this splits into partial tractions supported by the individual constituents on the parts of the domain where traction is specified. Similarly, while we might know the displacement or velocity of the mixture on parts of the boundary but not the individual displacements associated with the constituents (see Rajagopal and Tao [33] for detailed discussion of this issue). When one is dealing with a solid that swells significantly when infused with a fluid, like a sponge absorbing water or polymers and biological matter that undergo significant dimensional changes due to absorbing water, one can appeal to a variety of boundary conditions: (1) A boundary condition that stems from assuming that the boundary of the swollen solid that is in contact with the fluid is saturated (see Rajagopal et al. [34]); (2) Splitting the traction based on purely mechanical considerations (see Rajagopal and Tao [33]); (3) The assumption that the chemical potential is continuous across the boundary between the swollen solid and the fluid. When dealing with a mixture of ice and water, splitting the traction based the mass or volume fraction of the two constituents may be the best option.

10 Simplified Equations

We shall assume that there are only two constituents (the two phases: water and ice) in the mixture and there is the possibility of conversion between the phases. Furthermore, as a first step at studying the problem, we shall ignore the energy equation and study just the system of equations comprising those of mass and momentum balance for both the constituents, and even this simplified system reduces to eight coupled partial differential equations. In a future study, we shall consider the full thermodynamic problem that includes the balance of energy for the two phases as well as the second law of thermodynamics, which will add two more partial differential equations and a differential inequality.

In this case, the balance of mass reduces to:

$$\frac{\partial \rho^W}{\partial t} + \text{div}(\rho^W \mathbf{v}^W) = m, \tag{29}$$

and

$$\frac{\partial \rho^I}{\partial t} + \text{div}(\rho^I \mathbf{v}^I) = -m, \tag{30}$$

where m denotes the conversion of ice into water. Since we consider only two constituents, the production of one of the constituents is at the expense of the reduction of the other and hence $m^W = -m^I = m$. Next, we will assume that the partial stress tensors associated with both the phases is symmetric. Then, the balances of linear momenta reduce to:

$$\text{div}(\mathbf{T}^W) + \rho^W \mathbf{b}^W + \mathbf{m}^W + m\mathbf{v}^W = \frac{\partial}{\partial t}(\rho^W \mathbf{v}^W) + \text{div}(\rho^W \mathbf{v}^W \otimes \mathbf{v}^W), \tag{31}$$

and

$$\text{div}(\mathbf{T}^I) + \rho^I \mathbf{b}^I + \mathbf{m}^I - m\mathbf{v}^I = \frac{\partial}{\partial t}(\rho^I \mathbf{v}^I) + \text{div}(\rho^I \mathbf{v}^I \otimes \mathbf{v}^I). \tag{32}$$

We shall assume that

$$\mathbf{m}^W + \mathbf{m}^I + m(\mathbf{v}^W - \mathbf{v}^I) = 0. \tag{33}$$

We shall also assume that the body force acting on both the constituents is gravity, which we will denote by $\mathbf{g}$. Thus, the above equations reduce to

$$\mathrm{div}(\mathbf{T}^W) + \rho^W \mathbf{g} + \mathbf{m}^W + m\mathbf{v}^W = \frac{\partial}{\partial t}(\rho^W \mathbf{v}^W) + \mathrm{div}(\rho^W \mathbf{v}^W \otimes \mathbf{v}^W), \qquad (34)$$

$$\mathrm{div}(\mathbf{T}^I) + \rho^I \mathbf{g} + \mathbf{m}^I - m\mathbf{v}^I = \frac{\partial}{\partial t}(\rho^I \mathbf{v}^I) + \mathrm{div}(\rho^I \mathbf{v}^I \otimes \mathbf{v}^I), \qquad (35)$$

where $\mathbf{g}$ is the acceleration due to gravity. Since we have assumed that the stress tensors are symmetric, it follows from the balance of angular momentum that there can be no angular momentum supply to either of the constituents. The governing equations for the specific model under consideration are obtained by substituting the appropriate constitutive expressions for the production of water m, the interaction force $\mathbf{m}$ (discussed in Sect. 12) and the constitutive assumptions for the partial stresses for water and ice, $\mathbf{T}^w$ and $\mathbf{T}^I$, respectively.

11   Constitutive Relations: Model for Water and Ice

We shall model water as a Navier–Stokes fluid whose viscosity depends on temperature. Thus, we shall model the Cauchy stress in water through

$$\mathbf{T}^W = -p_{\mathrm{thm}}(\rho^W, \theta)\mathbf{I} + \lambda^W(\rho^W, \theta)\mathrm{tr}(\mathbf{A}_1^W)\mathbf{I} + \mu^W(\rho^W, \theta)\mathbf{A}_1^W, \qquad (36)$$

where $\mathbf{I}$ is the unit tensor, p_{thm} denotes the thermodynamic pressure of water that is assumed to be compressible, albeit slightly, in the temperature range of interest, ρ^W is the density of water, λ^W and μ^W are the bulk and shear viscosities of water, θ is the absolute temperature, and

$$\mathbf{A}_1^W = \frac{\partial \mathbf{v}^W}{\partial \mathbf{x}} + \left(\frac{\partial \mathbf{v}^W}{\partial \mathbf{x}}\right)^{\mathrm{T}} = 2\mathbf{D}^{\mathbf{W}}. \qquad (37)$$

The constitutive relation for ice depends on the time scale of interest. If one is concerned with long term response of ice sheets, they behave as though they are fluid-like. The popular model for ice is that which was proposed by Glen [10] to characterize the response of ice, as a power-law fluid model. But this model is incapable of describing normal stress differences in simple shear flow that has been observed in ice (see Kjartson et al. [18], Man and Sun [19]). A simple model that allows for the shear thinning as well as the normal stress differences observed in ice is the following:

$$\mathbf{T}^I = -p\mathbf{I} + \mu^I(\mathbf{A}_1^I, \theta)\mathbf{A}_1^I + \alpha_1 \mathbf{A}_2^I + \alpha_2 \mathbf{A}_1^I, \qquad (38)$$

where μ^I is the viscosity of ice, and α_1 and α_2 are the normal stress coefficients (which for the sake of simplicity we shall assume to be constant), and

$$\mathbf{A}_2^I = \frac{d\mathbf{A}_1^I}{dt} + \mathbf{A}_1^I \mathbf{L}^I + (\mathbf{L}^I)^{\mathrm{T}} \mathbf{A}_1^I, \tag{39}$$

where $\frac{d}{dt}$ is the material time derivative (see (9)), and $\mathbf{A}_1^I$ and $\mathbf{A}_2^I$ are the first two Rivlin–Ericksen tensors (see Rivlin and Ericksen [35]). We shall assume a power-law form for the generalized viscosity, namely

$$\mu^I(\mathbf{A}_1^I, \theta) = \hat{\mu}(\theta) \left[1 + \lambda (\mathrm{tr}(\mathbf{A}_1^I)^2) \right]^n, \tag{40}$$

where n is the power-law exponent and λ is a constant.

A critical and detailed discussion of the fluids of the differential type, the class to which the above model belongs, can be found in the review article by Dunn and Rajagopal [8]. The above model does not allow for ice to exhibit stress-relaxation. Stress-relaxation does not seem to be an important characteristic of ice and hence we shall use the simple model (38). If one requires a model that can also exhibit stress relaxation, then one could use a generalized rate type model that can describe shear thinning, stress-relaxation, nonlinear creep and normal stress differences in simple shear flow, namely a modified Maxwell model:

$$\mathbf{T}^I = -p^I \mathbf{I} + \mathbf{S}^I, \tag{41}$$

$$\mathbf{S}^I + \lambda \overset{\triangledown}{\mathbf{S}^I} = \eta(\mathbf{A}_1^I) \mathbf{A}_1^I, \tag{42}$$

where $-p^I \mathbf{I}$ is the indeterminate part of the stress due to the constraint of incompressibility, λ is the relaxation time, $\eta(\mathbf{A}_1^I)$ is the generalized viscosity and the upper convected Oldroyd derivative $\overset{\triangledown}{\mathbf{B}}$ is defined through

$$\overset{\triangledown}{\mathbf{B}} = \frac{d\mathbf{B}}{dt} - \mathbf{LB} - \mathbf{BL}^{\mathrm{T}}, \tag{43}$$

for any tensor $\mathbf{B}$.

12 Constitutive Assumption for the Interaction Terms

A key aspect of the constitutive theory for mixtures is the specification of interaction forces $\mathbf{m}^\alpha$ which can be used to incorporate the effect of drag due to the difference in velocity, the virtual mass effect, Magnus effect due to the relative spin, the effect of relative history of motion, the differences in the densities, etc., all of them between

the various constituents, (see Johnson et al. [16] for a detailed discussion of the
interaction mechanisms). In this work, in addition to taking into account the mass
that is generated due to phase change, we have to take into account interaction terms
that come into play, both in the balance of linear momentum and the balance of
energy due to phase change. With regard to the interaction terms that come into
play in the balance of linear momentum, as the flows involved are reasonably slow,
we can ignore effects such as the virtual mass effect that are a consequence of the
difference in the acceleration of the constituents. We shall also ignore the effects of
relative spin, relative histories, etc. The only interaction terms that we shall take into
account are due to the relative velocities, namely the Drag, and the interaction term
that is a consequence of the phase change. Thus, we shall assume that

$$\mathbf{m}^W = \alpha(\theta)(\mathbf{v}^W - \mathbf{v}^I) + \lambda^{IW}, \tag{44}$$

where α is the Drag coefficient and λ^{IW} denotes the contribution to the momentum
of water due to phase change. Similarly

$$\mathbf{m}^I = \alpha(\theta)(\mathbf{v}^I - \mathbf{v}^W) + \lambda^{WI}, \tag{45}$$

and

$$\lambda^{WI} \neq -\lambda^{IW}. \tag{46}$$

Specific constitutive choices have to be made for λ^{IW}, λ^{WI} and m^α and these
should be based on observations and carefully carried out experiments. We can
start by assuming simple forms such as those used by Morland et al. [25]. In a
comprehensive study that also includes the energy equation and the second law of
thermodynamics, we would have to specify constitutive relations for the internal
energy of each of the constituents, the energy supply ϵ^α, heat flux vector associated
with each constituent $\mathbf{q}^\alpha$, and the radiant energy supply r^α.

13   Governing Equations

Based on the above balance laws, the second law of thermodynamics, the volume
additivity constraint and constitutive relations, we have to obtain the governing
partial differential equations, and the inequality that stems from the second law.
Further simplifications can then be made that apply to the problem on hand. For
instance, as we shall be interested in the sea ice flow in the Arctic, we expect the
flow to be slow and thus we can neglect the inertial term in the balance of linear
momentum for the constituents. Also, as mentioned earlier, we can associate a
single temperature with both the constituents, water and ice. In this short paper
we are merely interested in documenting the basic balance laws and the constitutive

relations that will be used to develop the governing equations. In future work we will present the mathematical analysis of specific simplified models.

Acknowledgement KRR thanks the Office of Naval Research for support of this work.

References

1. Atkin, R.J., Craine, R.E.: Continuum theories of mixtures: basic theory and historical development. Q. J. Mech. Appl. Math. **29**, 209–244 (1976)
2. Atkin, R.J., Craine, R.E.: Theories of mixtures: applications. J. Inst. Math. Appl. **17**, 153–207 (1976)
3. Bedford, A., Drumheller, D.S.: Theory of immiscible and structured mixtures. Int. J. Eng. Sci. **21**, 863–960 (1983)
4. Bowen, R.: In: Eringen, A.C. (ed.) Theory of Mixtures in Continuum Physics III. Academic, New York (1976)
5. Bushuk, M., Giannakis, D., Majda, A.J.: Arctic sea ice reemergence: the role of large-scale oceanic and atmospheric variability. J. Climate **28**, 5477–5509 (2015)
6. Coleman, B.D., Noll, W.: An approximation theorem for functionals, with applications in continuum mechanics. Arch. Ration. Mech. Anal. **6**, 355–370 (1960)
7. Darcy, W.: Les Fontaines Publiques de La Ville de Dijon. Dalmont, Paris (1856)
8. Dunn, J.E., Rajagopal, K.R.: Fluids of differential type: critical review and thermodynamic analysis. Int. J. Eng. Sci. **33**, 689–729 (1995)
9. Fick, A.: Ueber diffusion. Ann. Phys. **94**, 59–86 (1855)
10. Glen, J.W.: The flow law of ice: a discussion of assumptions made in glacier theory, their experimental foundations and consequences, in the Physics of the Movement of Ice. Int. J. Assoc. Sci. Hydrol. Publ. **47**, 171–183 (1958)
11. Gray, J.M.N.T., Morland, L.W.: A two-dimensional model for the dynamics of sea ice. Philos. Trans. Phys. Sci. Eng. **347**(1682), 219–290 (1994)
12. Green, A.E., Naghdi, P.M.: On basic equations for mixtures. Q. J. Mech. Appl. Math. **XXII**, 427–438 (1969)
13. Hunke, E.C., Dukowicz, J.K.: An elastic-viscous-plastic model for sea ice dynamics. J. Phys. Occanogr. **27**, 1849–1867 (1997)
14. Hutter, K.: Theoretical Glaciology: Material Science of Ice and the Mechanics of Glaciers and Ice Sheets. D. Reidell, Norwell (1983)
15. Hutter, K., Blatter, H., Funk, M.: A model computation of moisture content in polythermal ice sheets. J. Geophys. Res. **93**, 12205–12214 (1988)
16. Johnson, G., Massoudi, M., Rajagopal, K.R.: A review of interaction mechanisms in fluid-solid flows, DOE/PETCLTR90/9, DE 91 0000941, Pittsburgh, PA (1991)
17. Kelly, R.J., Morland, L., Morris, E.M.: A three phase mixture model for melting snow. In: Modelling Snowmelt-induced Processes, vol. 155, pp. 17–26. IAHS Publications, Wallingford (1986)
18. Kjartanson, B.H., Shields, D.H., Domaschuk, L., Man, C.S.: The creep of ice measured with the pressuremeter. Can. Geotech. J. **25**, 250–261 (1988)
19. Man, C.S., Sun, Q.X.: On the significance of normal stress effects in the flow of glaciers. J. Glaciol. **33**, 268–273 (1987)
20. Marshall, S.J., Clarke, G.K.C.: Sensitivity analyses of coupled ice sheet/ice stream dynamics on the EISMINT experimental ice block. Ann. Geol. **23**, 336–347 (1996)
21. Marshall, S.J., Clarke, G.K.C.: A continuum mixture model for ice stream thermomechanics in the Laurentide ice sheet. J. Geophys. Res. **102**, 20599–20613 (1997)
22. Mills, N.: Incompressible mixtures of newtonian fluids. Int. J. Eng. Sci. **4**, 97–112 (1966)

23. Morland, L.: A simple constitutive theory for fluid saturated porous solid. J. Geophys. Res. **77**, 890–900 (1972)
24. Morland, L.: A theory of slow fluid flow through a thermoelastic porous solid. Geophys. J. Res. Astron. Soc. **55**, 393–410 (1978)
25. Morland, L., Kelly, R.J., Morris, E.M.: A mixture theory for phase-changing snowpack. Cold Reg. Sci. Technol. **17**, 271–285 (1990)
26. Morris, E.M.: Modelling the flow of mass and energy within a snowpack for hydrological forecasting. Ann. Glaciol. **4**, 198–203 (1986)
27. Perovich, D.K., Richter-Menge, J.R.: Loss of sea ice in the Arctic. Annu. Rev. Marine Sci. **1**, 417–441 (2009)
28. Prasad, S.C., Rajagopal, K.R.: On the diffusion of fluids through solids undergoing large deformations. Math. Mech. Solids **11**, 91–105 (2006)
29. Rajagopal, K.R.: On an hierarchy of approximate models for flows of incompressible fluids through porous solids. Math. Methods Models Appl. Sci. **17**, 215–252 (2007)
30. Rajagopal, K.R., Srinivasa, A.R.: A thermodynamic framework for rate type fluid models. J. Non-Newtonian Fluid Mech. **88**, 207–227 (2000)
31. Rajagopal, K.R., Srinivasa, A.R.: Thermomechanics of materials that have multiple natural configurations: Part 1 Viscoelasticity and classical plasticity. Z. Angew. Math. Phys. **55**, 861–893 (2004)
32. Rajagopal, K.R., Srinivasa, A.R.: On the thermodynamics of fluids defined by implicit constitutive relations. Z. Angew. Math. Phys. **59**, 715–729 (2008)
33. Rajagopal, K.R., Tao, L.: Mechanics of Mixtures. World Scientific, Singapore (1995)
34. Rajagopal, K.R., Wineman, A.S., Gandhi, M.V.: On boundary conditions for a certain class of problems in mixture theory. Int. J. Eng. Sci. **24**, 1453–1463 (1986)
35. Rivlin, R.S., Ericksen, J.L.: Stress deformation relations for isotropic materials. J. Ration. Mech. Anal. **4**, 323–425 (1955)
36. Samohyl, I.: Thermodynamics of Irreversible processes in Fluid Mixtures. Teubner, Leipzig (1987)
37. Shi, J.J., Rajagopal, K.R., Wineman, A.S.: Application of the theory of interacting continua to the diffusion of a fluid through a non-linear elastic media. Int. J. Eng. Sci. **19**, 871–889 (1981)
38. Truesdell, C.: Sulla basi della termomeccanica, Accademia Nazionale dei Lincei, Rendiconti della Classe di Scienze Fisiche, Mathematiche e Naturali (8) **22**, 33–88 (1957)
39. Truesdell, C.: Sulla basi della termomeccanica, Accademia Nazionale dei Lincei, Rendiconti della Classe di Scienze Fisiche, Mathematiche e Naturali (8) **22**, 158–166 (1957)
40. Truesdell, C.: Rational Thermodynamics. Springer, New York (1984)
41. Truesdell, C., Toupin, R.: The classical field theories. In: Flugge, W. (ed.) Handbuch der Physik, vol. III. Springer, New York (1960)
42. Ziegler, H.: Some extremum principles in irreversible thermodynamics. In: Sneddon, I.N., Hill, R. (eds.) Progress in Solid Mechanics, vol. 4. North Holland Publishing Company, New York (1963)
43. Ziegler, H.: An Introduction to Thermodynamics. North-Holland Series in Applied Mathematics and Mechanics, 2nd edn. North-Holland, Amsterdam (1983)
44. Ziegler, H., Wehrli, C.: The derivation of constitutive equations from the free energy and the dissipation function. In: Wu, T.Y., Hutchinson, J.W. (eds.) Advances in Applied Mechanics, vol. 25, pp. 183–238. Academic Press, New York (1987)

Modelling Sea Ice and Melt Ponds Evolution: Sensitivity to Microscale Heat Transfer Mechanisms

Andrea Scagliarini, Enrico Calzavarini, Daniela Mansutti, and Federico Toschi

Abstract We present a mathematical model describing the evolution of sea ice and meltwater during summer. The system is described by two coupled partial differential equations for the ice thickness h and pond depth w fields. We test the sensitivity of the model to variations of parameters controlling fluid-dynamic processes at the pond level, namely the variation of turbulent heat flux with pond depth and the lateral melting of ice enclosing a pond. We observe that different heat flux scalings determine different rates of total surface ablations, while the system is relatively robust in terms of probability distributions of pond surface areas. Finally, we study pond morphology in terms of fractal dimensions, showing that the role of lateral melting is minor, whereas there is evidence of an impact from the initial sea ice topography.

Keywords Glaciology · Sea ice · Turbulent heat transfer · Mathematical Modelling

A. Scagliarini (✉) · D. Mansutti
Istituto per le Applicazioni del Calcolo 'M. Picone', CNR, Rome, Italy
e-mail: andrea.scagliarini@cnr.it; d.mansutti@iac.cnr.it

E. Calzavarini
Université de Lille, Unité de Mécanique de Lille, UML EA 7512, Lille, France
e-mail: enrico.calzavarini@polytech-lille.fr

F. Toschi
Eindhoven University of Technology, Eindhoven, The Netherlands

Istituto per le Applicazioni del Calcolo 'M. Picone', CNR, Rome, Italy
e-mail: f.toschi@tue.nl

© Springer Nature Switzerland AG 2020
P. Cannarsa et al. (eds.), *Mathematical Approach to Climate Change and Its Impacts*, Springer INdAM Series 38,
https://doi.org/10.1007/978-3-030-38669-6_6

179

1 Introduction

The Arctic Ocean is characterised by the presence of ice, formed from the freezing of oceanic water. Such layer of sea ice is a key component of the Earth Climate System [1, 2], for it represents a sort of 'boundary condition' for heat, momentum and mass exchange between ocean and atmosphere at high latitudes [3–6] and plays a crucial role in the salinity balance in the ocean [7, 8], thus affecting also the thermohaline circulation [9]. Moreover, sea ice turns out to be a sensitive indicator of climate change: during the last few decades its average thickness and extent decreased significantly [10–12]. This decrease is two-way coupled with global warming, which shows up particularly striking in the Arctic, via the so called ice-albedo feedback. Sea ice, in fact, has a large albedo as compared to open oceanic waters, i.e. it reflects a high fraction of the incident solar radiation, while water absorbs it, thus favouring warming. The warmer the Earth surface the more ice melts, the lower gets the global albedo. The variability of sea ice emerges as the result of many processes acting on different time scales: the energy budget involving incoming and outgoing radiation [13–15], the melting phase transition [16, 17], the transport of water through ice porous structure [18–21], the rheology of internal stresses [22–25], the transport forced by couplings with ocean and atmosphere [26–31]. All these make sea ice an extremely complex system and its theoretical modelling a challenge [1, 2, 32, 33].

An important role in the ice-albedo feedback is played by the presence, on the ice surface, of melt ponds [34, 35]: during summer both the snow cover and the upper surface of sea ice melt and, as a consequence, meltwater may accumulate in depressions of the ice topography (thus forming ponds). The albedo of a melt pond ranges between ∼0.1 and ∼0.5 [36], while that of ice between ∼0.4 and ∼0.8 [34]. The average albedo for ponded ice is, then, lower than for the unponded one [37]. The evolution of melt ponds and of their distribution over the sea ice surface is, therefore, a key ingredient to be accounted for in realistic models of sea ice. It has been indeed suggested that a missing or improper inclusion of melt ponds could be the cause of overestimation, by certain general circulation models (GCMs), of the September sea ice minimum [38, 39]. For climatological temporal scales, it is important to get an accurate enough knowledge of the pond depth and surface area distributions, since these ones impact on the radiation budget; the rate of heat transfer through the ice pack, moreover, depends on the dynamics of meltwater, which, despite the average shallowness of ponds, can be turbulent [40].

The complexity of the melt-pond-covered sea ice system resides exactly in this intrinsic multiscale nature. Borrowing terms from Condensed Matter Physics, one can say that a modellistic approach may be tackled, at least, at three level of description: a *microscopic* level, where the focus is on the "atoms" of the system, the single pond and the fluid dynamics inside it, as done in, e.g., [41, 42]; a *mesoscopic* level, where the evolution of many ponds is considered, coupled with the evolution of a resolved sea ice topography [43–46]; and, finally, a *macroscopic* level, on scales of climatological interest, where sea ice dynamics is described in terms of an ice thickness distribution (ITD) [47–49], and melt ponds need to be parametrized [38, 50, 51]. The aim of this contribution is twofold. We will propose a *mesoscopic* model (in the sense explained above) and employ it to assess the sensitivity of the melt-ponds-covered sea ice system to different modelling of certain dynamical processes occurring at the single pond *microscopic* level.

The paper is organized as follows: in Sect. 2 we introduce the proposed mathematical model and its numerical implementation; in Sect. 3 the main results are illustrated and discussed, while concluding remarks and research outlooks are left to Sect. 4.

2 The Mathematical Model

The physical processes that occur within the ice pack and lead to variation of the sea ice thickness, can be grouped essentially into two categories: thermodynamic and mechanical. Thermodynamic processes are those related to the radiative budget; the fraction of incoming radiation that is absorbed is spent to increase the surface temperature and to melt ice. Mechanical deformations of sea ice are induced by ocean and wind stresses. These can drive sea ice transport, as well as elasto-plastic deformations in the pack, giving rise to events such as ridging and rafting [1]. Since we are interested in simulating processes involving ice melting and meltwater dynamics, we will neglect sea ice transport and mechanical terms (despite they can act on time scales comparable to melting in summer). As ice melts, meltwater is formed and transported, by sliding over the ice topography and seepage through its porous structure. It will eventually concentrate in *local minima* of the ice topography, forming melt ponds.

2.1 The Sea-Ice-Thickness/Melt-Pond-Depth System

We consider, therefore, the evolution of the ice (of density ρ_i) thickness field $h(\mathbf{x}, t) \geq 0$ and the meltwater (of density ρ_w) pond depth field $w(\mathbf{x}, t) \geq 0$ (with $\mathbf{x} \in \Omega \subset R^2$), whose dynamical equations read:

$$\partial_t h = -f \tag{1}$$

$$\partial_t w = -\nabla \cdot (\mathbf{u}w) + \frac{\rho_i}{\rho_w} f - s,$$

where f, $\mathbf{u}w$ and s represent, respectively, the melting rate, the meltwater flux (per unit cross-sectional area) and the seepage rate, which are, in general, functionals of h and w.

Similar mesoscopic models based on the evolution of h and w have been proposed in the past [43, 46]. Here, the original contributions to the modelling are in the parametrization of fluid-dynamics processes, in particular the water transport term and, more importantly, the vertical and lateral melt-rate term in turbulent flow conditions, which we will describe in detail in the following.

2.1.1 Melting Rate

The precise description of the energy budget at the sea ice cover, involving incoming and outgoing radiations and the thermodynamics of ice, can be quite a challenging task [13–15]. Being the focus of our study, though, a particular aspect of the melting process, namely the reduced albedo by meltwater covering the sea ice surface, we adopt a simple modelling [43], that proves, on the other hand, to be suitable to straightforward generalizations for the problems of interest here. We write the total melting rate f appearing in (1) as the sum of two terms

$$f = (1 - \chi)\phi_1(w) + \chi\phi_2(w, \nabla w, \nabla h); \qquad (2)$$

here, the first term, ϕ_1, is *local*, in fact it depends only on the pond depth $w(\mathbf{x}, t)$, whereas the second term, ϕ_2, includes also *lateral melting* mechanisms and may, thus, in principle depend also on gradients of the pond depth and ice thickness fields. The binary variable $\chi \in \{0, 1\}$ has been introduced to switch on ($\chi = 1$) or off ($\chi = 0$) such lateral melting contribution. Let us first discuss the local term ϕ_1. We assume a constant melting rate $\phi_1 = m_i$, of dimensions [length/time], for *bare* (unponded) ice (i.e. if $w(\mathbf{x}, t) = 0$), which is magnified by a w-dependent factor $\mathscr{A}(w)$, if ice is covered by a pond ($w(\mathbf{x}, t) > 0$); altogether, the expression for ϕ_1 reads:

$$\phi_1(w) = \mathscr{A}(w)m_i. \qquad (3)$$

Following Lüthje et al. [43], one can take $\mathscr{A}(w)$ to be:

$$\mathscr{A}(w) = \begin{cases} 1 + \dfrac{m_p}{m_i}\dfrac{w}{w_{\max}} & \text{if } w \in [0, w_{\max}] \\[2ex] 1 + \dfrac{m_p}{m_i} & \text{otherwise} \end{cases} \qquad (4)$$

where m_p is a (constant) limit melting rate for ponded ice, when the overlying pond depth exceeds the value $w_{\max}$ (which is usually estimated to be pretty small, $w_{\max} \approx 0.1\,\mathrm{m}$, because turbulent convection is already relevant at such depth, as discussed later on). The meaning and origin of such magnifying factor deserves some comments. In very shallow ponds, $w < w_{max}$, as a consequence of the absorption of solar radiation by water, the warming up is proportional to its volume and so the heat flux through the liquid layer is proportional to w. The situation changes for slightly deeper ponds, $w > w_{max}$, due to the appearance of natural convection. Indeed in summertime the temperature of air in contact with ponds

($\approx 2\,°C$) is higher than the basal one, in contact with melting ice (at $0\,°C$). In this range water density shows the well known anomaly, according to which it decreases with temperature, $\rho_w(T = 2\,°C) > \rho_w(T = 0\,°C)$, therefore, the pond is prone to convection. The latter sets on when the system becomes dynamically unstable; this will occur when the pond depth, which grows in time because of melting (thus making the system intrinsically non-stationary), will reach a value such that the time-dependent Rayleigh number $Ra(t)$ is large enough. The Rayleigh number quantifies the relative magnitude of buoyancy and dissipative terms; grouping together water density ρ_w, thermal expansion coefficient β, dynamic viscosity η, thermal conductivity κ and specific heat capacity at constant pressure c_p with gravity yields:

$$Ra(t) = \frac{c_p \rho_w^2 \beta g (\Delta T) w(t)^3}{\kappa \eta}, \tag{5}$$

Although it may seem surprising, the ponds being in general shallow, if we plug typical values in (5) we get, even for $w \approx 0.1\,\text{m}$ and $\Delta T \approx 0.2\,°C$, $Ra \approx 10^6$ [40], a value at which convection is already moderately turbulent [52]. Within ponds of depth $w \gtrsim 0.1\,\text{m}$, filled of fresh water, heat is not transferred by conduction, but by turbulent convection, whence the larger basal melting rate (3)–(4). For the sake of simplicity we neglect here salt concentration. Such an assumption must be taken with due care, though, since salinity hinders convection, by density stratification, and can even inhibit it (as shown in [53]).

The dependence of the total heat flux in turbulent conditions Φ_{turb} (in W/m^{-2} units) on the depth, though, is a complex problem. Expressed in non-dimensional variables, it amounts to assessing the Nusselt Nu vs Rayleigh numbers scaling $Nu \sim Ra^c$ [52, 54], where the Nusselt number is defined as:

$$Nu(t) = \frac{\Phi_{\text{turb}}(t)}{\kappa \frac{(\Delta T)}{w(t)}}. \tag{6}$$

The expression (4) arises from the assumption of the so called Malkus scaling $Nu \sim Ra^{1/3}$ [55]. Note that this scaling corresponds to the conjecture that the turbulent heat flux is independent of the thickness of the liquid layer, and as a consequence that the melt rate is fixed at a constant value m_p as stated by (4) in the model by Lüthje et al. [43] or by Taylor and Feltham [40]. However, theories, experiments and numerical simulations tend to agree that, in the range of Ra of relevance for melt pond convection, the scaling exponent should be $c < 1/3$ (see, e.g., [54] and references therein); in particular, widely observed is $Nu \sim Ra^c$, with $c \approx 2/7$. A similar scaling was observed, in numerical simulations, also for turbulent thermal convection with phase transition, where a boundary evolves, driven by melting [42], a setup which more closely resembles what occurs inside a melt pond. So, we propose to generalize Eqs. (3)–(4) for the local magnitude of melting to a generic $Nu \sim Ra^c$ relation and we obtain:

$$\phi_1(w) = m_i + m_p(w, c) \left(\frac{w}{w_{\max}}\right)^\alpha, \quad \text{with } \alpha = \begin{cases} 1 & \text{if } w \in [0, w_{\max}] \\ \\ 3c - 1 & \text{if } w > w_{\max}, \end{cases}$$

(7)

so that for ponds deeper than w_{max} Lüthje et al.'s case [43] is recovered for $\alpha = 0$, while scaling exponent equal to 2/7 yields for $\alpha = -1/7$. Notice that we have allowed also the constant m_p to be depth dependent in our model, $m_p \to m_p(w)$. This is done in order to include another aspect of realistic convection in Arctic ponds: the effect of a surface wind shear. At high latitudes, in fact, strong wind shear from the atmospheric boundary layer is present that can affect significantly sea ice dynamics (e.g. in the formation of sea-ice bridges [56]). Artic winds act on pond surfaces and are able, in principle, to strongly modify the convection patterns [41]. In such situations, turbulent heat flux is initially depleted, due to thermal plumes distortion by the shear [57, 58], and then it increases again, when turbulent forced convection becomes the dominant mechanism. On the line of the same arguments exposed in [58], based on Prandtl's mixing length theory [59], an expression for the coefficient $m_p(w)$ of the following form

$$m_p(w, c) \sim m_p^{(0)}(w, c) \left(\frac{a_1}{1 + c_1(\tau_s)w^{\gamma_1}} + a_2 c_2(\tau_s)w^{\gamma_2}\right),$$

(8)

can be expected, where a_1 and a_2 are some phenomenological parameters and c_1 and c_2 are functions of the wind shear magnitude τ_s (and of physical properties of meltwater). In all numerical results reported here, however, we have set $\tau_s = 0$, that is we have kept $m_p(w) \equiv m_p^{(0)}(w, c)$ (exploring wind shear effects will be object of a forthcoming study). The dependence of $m_p^{(0)}(w, c)$ on w and c stems from the fact that: (i) below w_{max} the heating is mainly radiative and (ii) changing the exponent of the scaling relation between dimensionless quantities, $Nu \sim Ra^c$, affects also the prefactor of the turbulent heat flux, i.e. $\Phi_{\text{turb}} = A(c)w^{3c-1}$. The expression for $m_p^{(0)}(w, c)$ therefore reads:

$$m_p^{(0)}(w, c) = \begin{cases} m_{p,r}^{(0)} & \text{if } w \in [0, w_{\max}] \\ \\ b_c(Pr) \left(\frac{c_p \rho_w^2 \beta g}{\eta}\right)^c \kappa^{1-c}(\Delta T)^{1+c} & \text{if } w > w_{\max}, \end{cases}$$

(9)

where the coefficient $b_c(Pr)$ depends on the Prandtl number, $Pr = c_p \eta / \kappa$. As previously commented, Eq. (7) is purely local and "vertical", in the sense that, if we think in discrete time, in a step Δt, it would increase the pond depth by $\phi_1(w(\mathbf{x}, t))\Delta t$, $w(\mathbf{x}, t) \to w(\mathbf{x}, t) + \frac{\rho_i}{\rho_w}\phi_1(w(\mathbf{x}, t), t)\Delta t$, and decrease the ice thickness by $h(\mathbf{x}, t) \to h(\mathbf{x}, t) - \phi_1(w(\mathbf{x}, t))\Delta t$, without affecting or being affected by the neighbourhood. We may expect, though, that, due to convection induced mixing, meltwater will be at a higher temperature than the surrounding ice and it may, therefore, favour melting also horizontally. This can be especially relevant close to the edge of pond surfaces, where it should give rise to a widening of ponds. To account for this kind of mechanism, we have introduced in the expression for the total melting rate, Eq. (2), the term $\phi_2(w, \nabla w, \nabla h)$, which contains the lateral melting (its explicit lattice expression will be given in Sect. 2.2). An attempt to estimate lateral fluxes in pond convection was proposed by Skyllingstad and Paulson [41], though with prescribed and fixed (with no evolving boundaries) forms of ponds. Finally, it is important to underline that by "lateral melting" we refer here to horizontal melting within the pond, and not edge melting of the ice pack, as when interactions with the ocean are considered [60].

2.1.2 Seepage Rate

Sea ice has a complex porous structure that evolve in time as the pack melts [19, 61]; a thorough description of water percolation through it is a formidable task that goes beyond the scope of the present work. We just model water transport through sea ice using Darcy's law; in addition, we distinguish between vertical and horizontal transport [43, 46]. Vertical transport is accounted for in Eqs. (1) by the seepage term s; the horizontal contribution, also dubbed lateral drainage, will be discussed in the next subsection. In order to derive an expression for the seepage rate, we recall that, according to Darcy's law, the discharge through volume of homogeneous porous material of permeability k, cross-sectional area a and length ℓ, under an applied pressure difference $(p_{\text{in}} - p_{\text{out}})$, is given by

$$q = k\frac{a(p_{\text{in}} - p_{\text{out}})}{\eta \ell};$$

(10)

for a portion of ponded ice of elementary area δa and thickness h, such pressure head is due to the hydrostatic pressure of the column of water in the pond overlying ice on δa, whose height is w, is $(p_{\text{in}} - p_{\text{out}}) = \rho_w g \delta a w$. The discharge q equals the time variation of the overlying volume of water, $\mathscr{V} = \delta a w$, providing

$$\dot{w} = -k\frac{\rho_w g w}{\eta h},$$

(11)

out of which we can read the expression for the seepage rate s that is [46]

$$s = k\frac{\rho_w g}{\eta}\frac{w}{h}.$$

(12)

2.1.3 Meltwater Flux

The seepage rate just introduced, Eq. (12), entails a dependence of the equation for $w(\mathbf{x}, t)$ on $h(\mathbf{x}, t)$ (that would be otherwise be decoupled from it, as far as only melting is concerned). A further coupling is induced by the transport term and the associated meltwater flux $\mathbf{u}$. Such term is also the only non-local one in the evolution (for it involves derivatives of h and w), thus introducing a dependence of the dynamics on the ice topography. It represents, in other words, the driving for meltwater to accumulate to form ponds. The transport of meltwater is realised essentially with two mechanisms: *sliding* of water over slopes of the ice surface and *lateral drainage* through the porous structure of ice. Correspondingly, the flux consists of the sum of two terms

$$\mathbf{u} = \mathbf{u}_{\text{sliding}} + \mathbf{u}_{\text{drainage}}; \tag{13}$$

as discussed in the previous subsection, $\mathbf{u}_{\text{drainage}}$ stems from the horizontal component of Darcy's law and, hence, is given by [43]

$$\mathbf{u}_{\text{drainage}} = -\Pi \frac{\rho_w g}{\eta} \nabla (h + w), \tag{14}$$

where Π is the horizontal permeability of ice.

In order to model the sliding term, we resort to the theory of shallow water equations (SWE) [62], considering that the width of a layer of water sliding over the ice topography is relatively thin. If we assume, furthermore, that the Reynolds number is small (we expect so, and a consequent creeping flow, for a thin layer of water sliding over the ice topography, the thickening of such layer being inhibited by seepage), the SWE for the depth-averaged two-dimensional velocity field reduce to the following balance equation between stresses at the bottom (due to friction with ice) and top (induced by wind forcing) of the fluid layer and gravity [63, 64] (assuming a no-slip boundary condition between water and ice and neglecting capillary effects)

$$\frac{3\eta}{w} \mathbf{u}_{\text{sliding}} + \tau_s + g w \nabla (h + w) \approx 0, \tag{15}$$

which yields for $\mathbf{u}_{\text{sliding}}$:

$$\mathbf{u}_{\text{sliding}} = -\frac{g w^2}{3\eta} \nabla (h + w) + \frac{\tau_s w}{3\eta} \hat{\tau}_s, \tag{16}$$

where $\hat{\tau}_s$ is the direction of the wind shear vector at the free water surface and τ_s is its magnitude, as in Eq. (8). Let us stress that, in this way, we have introduced, through Eqs. (8) and (16) a first minimal coupling of the model for the sea-ice-melt-ponds system with the atmospheric dynamics.

2.2 Numerical Implementation

The system of equations (1) is solved by means of a finite differences scheme; upon discretization on a square $M \times M$ lattice, with $M = 1024$, of equally Δ-spaced nodes, the system is converted in a set of coupled ordinary differential equations for the variables $h_{ij}(t) \equiv h(x_i, y_j, t)$ (with $x_i = i\Delta$, $y_j = j\Delta$ and $i, j = 1, 2, \ldots, M$) and $w_{ij}(t) \equiv w(x_i, y_j, t)$, that are, then, integrated numerically using a standard explicit Runge-Kutta fourth order time marching scheme with time step $\Delta t = 60\,\mathrm{s}$, that allows to resolve the fastest time scales of the meltwater transport terms. Spatial derivatives are approximated by the corresponding second order accuracy central differences. The lattice spacing Δ is taken to be $\Delta = 1\,\mathrm{m}$, so the physical size of the simulated system is $L^2 \approx 1\,\mathrm{km}^2$, where $L = M\Delta$; this choice is dictated by the condition that Δ is $\Delta \gtrsim \sigma_h$ (σ_h being the standard deviation of the initial ice thickness distribution), such that no significant variations of of h occur within one lattice spacing, i.e. the spatial derivative is at most $h'(x) \sim 1$, assuming that the average finite height variation over a Δ is $\Delta h \propto \sigma_h$. Periodic boundary conditions apply, so we neglect edge effects, such as water run-off and direct coupling with the ocean (e.g. lateral melting of floe, ocean stresses), i.e. it is as if we were simulating a virtually infinite sea ice floe.

The melting term ϕ_2, appearing in Eq. (2), takes the following expression on the lattice

$$\phi_{2_{i,j}} = \phi_{2_{i,j}}^{(V)} + \sum_{i'=\pm 1} \phi_{2_{i+i',j}}^{(L,x)} \Theta(w_{i+i',j} - w_{i,j}) + \sum_{j'=\pm 1} \phi_{2_{i,j+j'}}^{(L,y)} \Theta(w_{i,j+j'} - w_{i,j}),$$

$$(17)$$

which contains a combination of *vertical*, $\phi_{2_{i,j}}^{(V)}$, and *lateral*, $\phi_{2_{i,j}}^{(L,(x,y))}$, components of the melting; the latter are given by:

$$\phi_{2_{i,j}}^{(V)} = \phi_{1_{i,j}} \frac{1}{\sqrt{1 + (\hat{\partial}_x w_{i,j})^2 + (\hat{\partial}_y w_{i,j})^2}} \tag{18}$$

and

$$\phi_{2,i,j}^{(L,(x,y))} = \phi_{1_{i,j}} \frac{|\hat{\partial}_{(x,y)} w_{i,j}|}{\sqrt{1 + (\hat{\partial}_x w_{i,j})^2 + (\hat{\partial}_y w_{i,j})^2}}, \tag{19}$$

where $\hat{\partial}_{(x,y)}$ stands for the finite difference derivative. We assume that the magnitude of the turbulent heat flux is homogeneously distributed over the pond walls (that is at the ice/water interface) and its direction is parallel to the normal $\hat{n}$ to the interface. Therefore, the vertical and lateral contributions to the melting rate are weighted with the absolute values of the components of $\hat{n}$,

$$\frac{1}{\sqrt{1 + (\hat{\partial}_x w_{i,j})^2 + (\hat{\partial}_y w_{i,j})^2}} \left(|\partial_x w|, |\partial_y w|, 1 \right),$$

whence Eqs. (18) and (19). In other words, this means that, for instance, at the bottom of the pond mostly the vertical term will act, while when the topography is steep, as, e.g., next to the pond edge, ice ablation will be dominated by lateral melting. The presence of the Heaviside's functions, Θ, in (17) is to guarantee that the, non-local, lateral contribution to melting on a given site comes only from those neighbours that have a larger amount of overlying water (larger w). This is motivated by the idea that, if at a given elevation H a certain site is in the 'ice state', it will get a lateral melting contribution from a neighbouring site which, at the same elevation, is in a 'water state', since melting is driven by water convection in contact with ice enclosing the pond.

3 Results

The initial values of the sea ice topography $h_{ij}^0 \equiv h(x_i, y_j, 0)$ are random Gaussian numbers with given mean and variance. The initial topography is spatially correlated over a characteristic length $\delta \approx 8$ m. Two types of ice are used as initial conditions, namely first-year ice (FYI) and multi-year ice (MYI). FYI is newly formed in the winter preceding the melt season and is typically flatter, whereas MYI, that has overcome one or more melt seasons, presents a more rugged surface profile, i.e. it is characterized by larger variance and mean as compared to FYI. Consequently, wide and ramified but shallow melt ponds are more probably formed on FYI, while melt ponds on MYI will be tendentially deeper, of limited areal extension and more regularly shaped [50]. The initial condition is therefore expected to play an important role on the meltwater dynamics. The statistical parameters (mean $\langle h \rangle$ and variance σ_h of the thickness distribution) employed are $\langle h \rangle = 0.92$ m, $\sigma_h = 0.18$ m, for FYI, and $\langle h \rangle = 3.67$ m, $\sigma_h = 1.5$ m, for MYI [43, 65]. Other numerical values for the model parameters, which are kept fixed in all simulations, are summarized in Table 1. Evidently, we are faced to a wide, multi-dimensional, parameter space; many of these parameters (such as permeabilities and melting rates) are known only with limited accuracy and the system can be quite sensitive to their values. A full sensitivity study in such sense is somehow beyond the scope of the present work; moreover some studies of this kind (on similar models) are available (see, e.g. [43, 46]). We limit here ourselves, therefore, to test the novelties of the present model, namely the melting rate exponent associated to turbulent thermal convection and its contribution along the lateral (horizontal) directions.

Table 1 Values of model parameters which are kept fixed in all simulations: water density ρ_w, ice density ρ_i, water dynamic viscosity η, acceleration of gravity g, horizontal permeability of ice Π, *bare* ice melting rate m_i, melting rate enhancement factor $m_p^{(0)}$ and critical pond depth for melting rate enhancement w_{max}

Parameter	ρ_w	ρ_i	η	g	Π	m_i	$m_p^{(0)}$	w_{max}
Units	kg/m^3	kg/m^3	kg/(m s^{-1})	m/s^2	m^2	cm/day	cm/day	m
Value	1000	950	1.79×10^{-3}	9.81	3×10^{-9}	1.2	2	0.1

Fig. 1 Configuration of the depth field $w(\mathbf{x}, t)$ showing the melt ponds distribution over the sea ice surface, for FYI after 20 simulated days (a 200×200 m^2 region at the centre of the simulated domain is taken). White color corresponds to bare ice and blue color indicates the presence of a pond, the darker the blue the deeper the pond (deepest ponds have $w \approx 2$ m) (Color figure online)

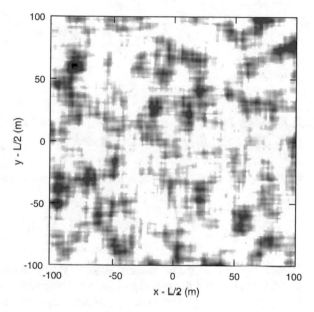

Snow cover is absent and no melt water is assumed at the initial time (i.e. $w(\mathbf{x}, t_0) = 0 \quad \forall \mathbf{x}$). As said before, we aim to simulate the summer time evolution of sea ice, so our t_0 is to be considered June 1st and, in view of this, refreezing of meltwater is not accounted for. We ran each simulation for ≈ 30 days. A visualization of the distribution of ponds corresponding to day 20 from the beginning of the simulation is shown in Fig. 1. In order to extract statistical information on the melt pond coverage of the sea ice, we first need to identify individual ponds. To do this, for each time t we define a pond as any connected subset of points on the lattice such that $w(\mathbf{x}, t) > 0$; the full pond configuration is determined by a cluster analysis (for which we employ the so called Hoshen-Kopelman algorithm [66]) over the whole system. The area of the i-th pond is then $A_i = n_i \Delta_x \Delta_y$, n_i being the number of points in the i-th cluster.

Fig. 2 Mean pond area vs
time for the 1/3 (red squares)
and 2/7 (blue circles) laws
(Color figure online)

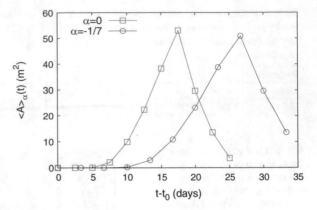

3.1 Melt Pond Areas Evolution: Role of the Turbulent Heat Flux Scaling Inside the Pond

In Fig. 2 we plot the time evolution of the mean pond area

$$\langle A \rangle_\alpha(t) = \frac{1}{N(t)} \sum_{i=1}^{N(t)} A_i(t) \tag{20}$$

(where $N(t)$ is the total of ponds detected at time t) for a FYI and assuming Malkus and 2/7 scaling for the turbulent heat flux, respectively, that is, with reference to Eq. (7), with $\alpha = 0$ (red squares, equivalent to the study in [43]) and $\alpha = -1/7$ (blue circles). The mean pond area grows and reaches a maximum faster when $\alpha = 0$: after 13 days, e.g., the 2/7-model gives a prediction for $\langle A \rangle_\alpha$ approximately seven times smaller than it is for the constant flux case; this suggests how an apparently minor assumption at the level of fluid dynamic processes within the single pond may lead to bad estimates on climatologically relevant indicators, such as the September sea ice extension. For the same two runs, with $\alpha = 0, -1/7$, we measured the probability distribution functions (PDFs) of pond areas, $P_\alpha(A, t)$, after 13 days; one can see from Fig. 3 that the two PDFs differ, although both seem to show a power law behaviour. Nevertheless, if we consider PDFs with *equal mean*, instead of *equal time* PDFs, interestingly, the two sets of points (for $\alpha = 0$ and $\alpha = -1/7$) collapse onto each other, as shown in Fig. 4. There we plot $P_{\alpha=0}(A, t_1)$ and $P_{\alpha=-1/7}(A, t_2)$, where t_1 and t_2 are such that $\langle A \rangle_{\alpha=0}(t_1) = \langle A \rangle_{\alpha=-1/7}(t_2)$; with reference to Fig. 2, this occurs, for instance, if we pick $t_1 - t_0 = 13$ days and $t_2 - t_0 = 20$ days, i.e. on June 14th for $\alpha = 0$ and June 21st for $\alpha = -1/7$. The two PDFs nicely follow the scaling $P_\alpha(A) \sim A^{-1.5}$ for relatively small areas ($A < 20\,\text{m}^2$), with

Fig. 3 Probability distribution functions of pond areas for the 1/3 and 2/7 laws at 14th June

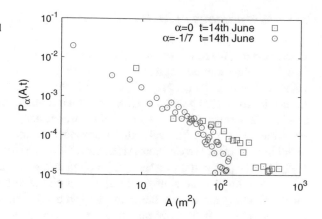

Fig. 4 PDFs of pond areas after 13, for $\alpha = 0$, and 20 days, for $\alpha = -1/7$: notice that the two sets of points basically overlap. The dashed and solid lines correspond to the power law $A^{-1.5}$ and $A^{-1.8}$, respectively

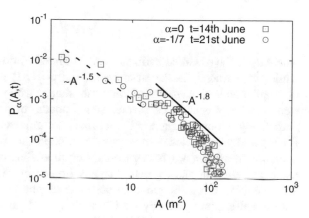

a steeper fall-off for larger values, $P_\alpha(A) \sim A^{-1.8}$. Such functional forms agree with available observational data, as those collected by means of aerial photography [35] during the SHEBA (Surface Heat Budget of the Arctic Ocean) [67, 68] and HOTRAX (*Healy-Oden* TRans Arctic EXpedition) [69] campaigns. Remarkably, the same power-laws for the PDFs were found in recent theoretical/numerical works based on statistical models (in the spirit of equilibrium statistical mechanics) [70, 71]. We would like to highlight, at this point, that this is a striking aspect of the melt-pond-sea-ice system: the melt pond system on large scales is robust with respect to area distribution (and pond geometry, as we shall see later on) against changes of certain physical parameters controlling the dynamics. So robust that even simple models, that do not account for the physics of the melt pond formation and evolution at all, can capture such statistical fingerprints. We will focus, then, on an aspect of melt pond configuration that one might expect to be affected by details of the evolution, namely their morphology.

3.2 Morphology of Melt Ponds: Role of Lateral Melting

Characterizing the morphology of the global ponds configuration and understanding
how it emerges can be of great relevance also for large scale models of sea ice (in
GCMs). There, in fact, a major limitation is due to the difficulty to relate properly
the sea ice topography with the redistribution of meltwater; ideally, one would wish
to know how much ice area is covered by water, and how deeply (since these two
quantities determine, basically, the absorbance of incident radiation). Analyzing
aerial images from two different Arctic expeditions, SHEBA [67, 68] and HOTRAX
[69], Hohenegger and coworkers [72] looked at the scatter plot of perimeter p and
area A of a multi-pond configuration; such a plot is known to contain information
on the fractal geometry of the manifold (embedded in a two-dimensional space)
considered [73, 74]. The two quantities are, in fact, related by

$$p \sim A^{d_p/2}, \tag{21}$$

where d_p is the so called perimeter fractal dimension: for a smooth curve $d_p = 1$,
while for a fractal, in the strict sense, $d_p > 1$. It was observed that surfaces
of small ponds tend to be of roundish shape, while large ones, that typically
stem from aggregation of several small ponds, display features of clusters in
percolating systems and appear fractal-like [72]. This transition to a fractal geometry
is supposedly connected with the way melt ponds grow over the sea ice surface; it
is natural to ask, then, whether the explicit modelling of the physical mechanisms
leading to such in-plane growth has any impact on the final global morphology.
This amounts to test the model in presence of what we called lateral melting, i.e.
with a melting rate given by Eq. (2), with $\chi = 1$, and (17). To this aim, we ran
the same simulation, for the FYI, as discussed above, with the lateral melting term
switched on. In Fig. 5 we show p vs A scatter plot after 20 days for two simulations
with (blue asterisks) and without (red bullets) lateral melting (symbols relative to

Fig. 5 Scatter plot of
perimeter and area for all
melt ponds on 21st June. The
dashed and solid lines
indicate the power law $A^{1/2}$
and A, corresponding to
perimeter fractal dimensions
$d_p = 1$ (smooth shapes) and
$d_p = 2$ (fractal), respectively;
the transition between the two
regimes occur at $A \approx 100\,\mathrm{m}^2$

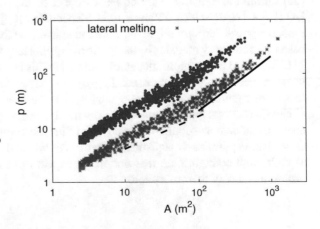

the two data sets are shifted from each other by a factor 3 for the sake of clarity, otherwise they would overlap). The two power laws, $p \sim A^{1/2}$, for $A < A_c$, and $p \sim A$, for $A > A_c$, are reasonably well followed in both cases; the only minor effect of the presence of lateral melting seems to be a slightly clearer scaling behaviour (especially for large A). The transition to the fractal geometry occurs at $A_c \approx 100 \, \mathrm{m^2}$, in agreement with the observations [72]. This same phenomenology was captured also by the above mentioned statistical models [70, 71], underlining further the robustness of the melt-pond-covered sea ice system as far as geometry is concerned.

To get a deeper insight on this aspect of melt pond configuration over the sea ice surface, we performed an analysis of the *generalized fractal dimensions* (GFD), or Rényi's q-entropies, $D(q)$ [75], which provide a more detailed description of the geometry of the fractal manifold. Let us call $\mu_i(\varepsilon)$ the measure of ponds within the i-th element of a regular tessellation of the domain in squares of side ε (i.e. the fractional area of the ε-square occupied by meltwater), and $N(\varepsilon)$ the total number of squares into which the domain is partitioned; we can then define the following quantity:

$$I(q, \varepsilon) = \frac{1}{1-q} \log \left(\sum_{i=1}^{N(\varepsilon)} \mu_i(\varepsilon)^q \right), \tag{22}$$

for any positive $q \neq 1$. The GFD are then computed as

$$D(q) = \lim_{\varepsilon \to 0} \frac{I(q, \varepsilon)}{\log(1/\varepsilon)}; \tag{23}$$

for $q = 0$, $I(0, \varepsilon)$ equals the number of non-void elements of the tessellation, therefore $D(q = 0) \equiv D_0$ coincides with the Haussdorf, or box-counting, dimension, which is an estimate of the fractal dimension of the set [73]. It is clear, then, in which sense the $D(q)$ are generalized fractal dimensions. For "ordinary" fractals, all the GFD are equal, i.e. $D(q)$ is constant with q. In general, though, it might be a non-increasing function of the order q: if this is the case, one talks about a *multifractal* set, that is a fractal whose dimension vary in space. In Fig. 6 we show the $D(q)$ computed from numerical data from three simulations, namely: FYI without lateral melting, FYI with lateral melting and MYI. The plot tells us that melt ponds on FYI lay on a fractal manifold, as suggested also by the perimeter-area relation, since $D_0 < 2$, whereas those on MYI do not $D(q) = 2 \; \forall q$; however, we observe a modest decrease of $D(q)$ with q, indicating a weak multifractality, with very minor (if any) differences between the run with lateral melting and the one without. These results are a first attempt to show that the morphology of the melt pond system can be even more complicated than what can be captured with the perimeter-area relations, which are known to give sometimes biased estimates of the actual fractal dimension for the areas [74].

Fig. 6 Generalized fractal (or Rényi's) dimensions for the melt pond distribution after 70 days from three simulations: FYI without lateral melting (red squares), FYI without lateral melting (blue dots) and MYI without lateral melting (green triangles) (Color figure online)

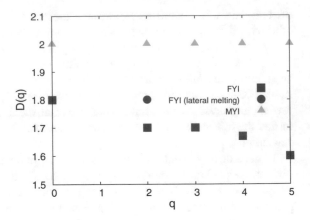

4 Conclusions and Perspectives

We have proposed a continuum *mesoscale* model that describes the evolution of Arctic sea ice, in presence of a coverage of meltwater ponds, that alter the sea ice thermodynamics (in terms of melting rates). The model consists of two coupled partial differential equations, for the ice thickness and pond depth fields. The physics of sea ice was kept at a very basic level in order to focus on the effect of dynamic processes occurring at the single pond level on the large scale configurations of the melt pond system and on sea ice evolution. Numerical simulations of the model showed that a minimal variation of the scaling exponent of the turbulent heat flux, within the pond, with the surface temperature impacts the time evolution of the mean pond size (shifting the maximum by few days), hence of the average melting rate. We stress that the assumption made in this work, that the melt rate in ponds (deeper than w_{max}) is a weakly decreasing function of the water layer depth rather than a constant is supported by a vast amount of studies on turbulent heat transfer. Therefore, our study suggests that a thorough knowledge and parametrization of melt pond hydrodynamics is needed in order to not get wrong estimates of observables of climatological relevance. On the other hand, statistical and geometrical properties of the melt pond system, such as the probability distribution function of pond surface areas and fractal dimensions, appeared to be robust against heat flux scaling variations as well as against the inclusion or not of an explicit modelling of lateral melting inside the pond. In particular, our results agreed well with observation for what concerns the power law decay of the PDFs and the perimeter-area relation, and the corresponding perimeter fractal dimension, for ponds. Finally, we have extended the study of melt ponds geometry to the analysis of generalized fractal dimensions, which showed a clear dependence on the initial ice topography, with melt ponds on first-year ice displaying even a weak multifractality, while those on multi-year ice being essentially smooth. This dependence on the initial condition suggests that, for future, studies, it would be of great interest to initialize the numerical model with conditions taken from field measurements. The study of the effect of wind stresses at the ice surface on global melting as well as on melt pond distribution and morphology will be a first extension of the present study. A further step forward that might be taken, within this approach, is the inclusion of a proper description of mechanical processes and rheology of sea ice, specially focusing on the effect of the presence of accumulation of meltwater on the local deformation properties of the pack.

Acknowledgements AS and DM acknowledge financial support from the National Group of Mathematical Physics of the Italian National Institute of High Mathematics (GNFM-INdAM). EC acknowledge supports form the French National Agency for Research (ANR) under the grant SEAS (ANR-13-JS09-0010).

References

1. Hunke, E.C., Lipscomb, W.H., Turner, A.K.: Sea-ice models for climate study: retrospective and new directions. J. Glaciol. **56**, 1162–1172 (2010)
2. Notz, D.: Challenges in simulating sea ice in Earth System Models. WIREs Clim. Change **3**, 509–526 (2012)
3. Cattle, H., Crossley, J.: Modelling Arctic climate change. Philos. Trans. R. Soc. A **352**, 1699 (1995)
4. Ebert, E.E., Schramm, J.L., Curry, J.A.: Disposition of solar radiation in sea ice and the upper ocean. J. Geophys. Res. **100**(C8), 965–975 (1995)
5. Maykut, G.A., McPhee, M.G.: Solar heating of the arctic mixed layer. J. Geophys. Res. **100**(C12), 24691–24703 (1995)
6. Tsamados, M., Feltham, D.L., Schröder, D., Flocco, D., Farrell, S.L., Kurtz, N., Laxon, S.W., Dacon, S.: Impact of variable atmospheric and oceanic form drag on simulations of arctic sea ice. J. Phys. Oceanogr. **44**, 1329–1353 (2014)
7. Vancoppenolle, M., Fichefet, T., Goosse, H., Bouillon, S., Madec, G., Morales Maqueda, M.A.: Simulating the mass balance and salinity of Arctic and Antarctic sea ice. 1. Model description and validation. Ocean Model. **27**, 33–53 (2009)
8. Vancoppenolle, M., Fichefet, T., Goosse, H.: Simulating the mass balance and salinity of Arctic and Antarctic sea ice. 2. Importance of sea ice salinity variations. Ocean Model. **27**, 54–69 (2009)
9. Mauritzen, C., Häkkinen, S.: Influence of sea ice on the thermohaline circulation in the Arctic-North Atlantic Ocean. Geophys. Res. Lett. **24**, 3257–3260 (1997)
10. Kwok, R., Cunningham, G.F., Wensnahan, M., Rigor, I., Zwally, H.J., Yi, D.: Thinning and volume loss of the Arctic Ocean sea ice cover: 2003–2008. J. Geophys. Res. **114**, C07005 (2009)
11. Stroeve, J.C., Kattsov, V., Barrett, A., Serreze, M., Pavlova, T., Holland, M., Meier, W.N.: Trends in Arctic sea ice extent from CMIP5, CMIP3 and observations. Geophys. Res. Lett. **39**, L16502 (2012)
12. Laxon, S.W., Giles, K.A., Ridout, A.L., Wingham, D.J., Willatt, R., Cullen, R., Kwok, R., Schweiger, A., Zhang, J., Haas, C., Hendricks, S., Krishfield, R., Kurtz, N., Farrell, S., Davidson, M.: CryoSat-2 estimates of Arctic sea ice thickness and volume, Geophys. Res. Lett. **40**, 732–737 (2013)
13. Maykut, G.A., Untersteiner, N.: Some results from a time-dependent thermodynamic model of sea ice. J. Geophys. Res. **76**(6), 1550–1575 (1971)

14. Ebert, E.E., Curry, J.A.: An intermediate one-dimensional thermodynamic sea ice model for investigating ice–atmosphere interactions, J. Geophys. Res. **98**(C6), 10085–10109 (1993)
15. Eisenman, I., Wettlaufer, J.S.: Nonlinear threshold behavior during the loss of Arctic sea ice. Proc. Natl. Acad. Sci. **106**, 28–32 (2009)
16. Steele, M.: Sea ice melting and floe geometry in a simple ice-ocean model. J. Geophys. Res. **97**(C11), 17729–17738 (1992)
17. Bitz, C.M., Lipscomb, W.H.: An energy-conserving thermodynamic model of sea ice. J. Geophys. Res. **104**(C7), 15669–15677 (1999)
18. Freitag, J., Eicken, H.: Melt water circulation and permeability of Arctic summer sea ice derived from hydrological field experiments. J. Glaciol. **49**, 349–358 (2003)
19. Feltham, D.L., Untersteiner, N., Wettlaufer, J.S., Worster, M.G.: Sea ice is a mushy layer. Geophys. Res. Lett. **33**, L14501 (2006)
20. Wells, A.J., Wettlaufer, J.S., Orszag, S.A.: Nonlinear mushy-layer convection with chimneys: stability and optimal solute fluxes. J. Fluid Mech. **716**, 203–227 (2013)
21. Turner, A.K., Hunke, E.C.: Impacts of a mushy-layer thermodynamic approach in global sea-ice simulations using the CICE sea-ice model. J. Geophys. Res. Oceans **120**(2), 1253–1275 (2015)
22. Feltham, D.L.: Sea ice rheology. Annu. Rev. Fluid Mech. **40**, 91–112 (2008)
23. Hunke, E.C., Dukowicz, J.K.: An elastic-viscous-plastic model for sea ice dynamics. J. Phys. Oceanogr. **27**(9), 1849–1867 (1997)
24. Tsamados, M., Feltham, D.L., Wilchinsky, A.: Impact of a new anisotropic rheology on simulations of Arctic sea ice. J. Geophys. Res. Oceans **118**(1), 91–107 (2013)
25. Rabatel, M., Rampal, P., Carrassi, A., Bertino, L., Jones, C.K.R.T.: Impact of rheology on probabilistic forecasts of sea ice trajectories: application for search and rescue operations in the Arctic. Cryosphere **12**, 935–953 (2018)
26. Steele, M., Zhang, J., Rothrock, D., Stern, H.: The force balance of sea ice in a numerical model of the Arctic Ocean. J. Geophys. Res. Oceans **102**(C9), 21061–21079 (1997)
27. Schröder, D., Vihma, T., Kerber, A., Brümmer, B.: On the parameterization of turbulent surface fluxes over heterogeneous sea ice surfaces. J. Geophys. Res. **108**(C6), 3195 (2003)
28. Rampal, P., Weiss, J., Marsan, D.: Positive trend in Arctic sea ice mean speed and deformation 1979–2007. J. Geophys. Res. Oceans **114**, C05013 (2009)
29. Rampal, P., Weiss, J., Dubois, C., Campin, J.-M.: IPCC climate models do not capture the Arctic sea ice drift acceleration: consequences in terms of projected sea ice thinning and decline. J. Geophys. Res. Oceans **116**, C00D07 (2011)
30. Petty, A.A., Feltham, D.L., Holland, P.R.: Impact of atmospheric forcing on Antarctic continental shelf water masses. J. Phys. Oceanogr. **43**(5), 920–940 (2013)
31. Heorton, H.D.B.S., Feltham, D.L., Tsamados, M.: Stress and deformation characteristics of sea ice in a high-resolution, anisotropic sea ice model. Philos. Trans. A Math. Phys. Eng. Sci. **376**(2129), 20170349 (2018)
32. Hunke, E.C., Notz, D., Turner, A.K., Vancoppenolle, M.: The multiphase physics of sea ice: a review for model developers. Cryosphere **5**, 989–1009 (2011)
33. Massonnet, F., Vancoppenolle, M., Goosse, H., Docquier, D., Fichfet, T., Blanchard-Wrigglesworth, E.: Arctic sea-ice change tied to its mean state through thermodynamic processes. Nat. Clim. Change **8**, 599–603 (2018)
34. Fetterer, F., Untersteiner, N.: Observations of melt ponds on Arctic sea ice. J. Geophys. Res. **103**(C11), 24821–24835 (1998)
35. Perovich, D.K., Tucker III, W.B., Ligett, K.A.: Aerial observations of the evolution of ice surface conditions during summer. J. Geophys. Res. **107**(C10), 8048 (2002)
36. Hanesiak, J.M., Barber, D.G., De Abreu, R.A., Yackel, J.J.: Local and regional albedo observations of arctic first-year sea ice during melt ponding. J. Geophys. Res. **106**(C1), 1005–1016 (2001)
37. Perovich, D.K., Grenfell, T.C., Light, B., Hobbs, P.V.: Seasonal evolution of the albedo of multiyear Arctic sea ice. J. Geophys. Res. **107**(C10), 8044 (2002)

38. Flocco, D., Schroeder, D., Feltham, D.L., Hunke, E.C.: Impact of melt ponds on Arctic sea ice simulations from 1990 to 2007. J. Geophys. Res. **117**, C09032 (2012)
39. Schröder, D., Feltham, D.L., Flocco, D., Tsamados, M.: September Arctic sea-ice minimum predicted by spring melt-pond fraction. Nat. Clim. Change **4**(5), 353–357 (2014)
40. Taylor, P.D., Feltham, D.L.: A model of melt pond evolution on sea ice. J. Geophys. Res. **109**, C12007 (2004)
41. Skyllingstad, E.D., Paulson, C.A.: A numerical simulations of melt ponds. J. Geophys. Res. **112**, C08015 (2007)
42. Rabbanipour Esfahani, B., Hirata, S.C., Berti, S., Calzavarini, E.: Basal melting driven by turbulent thermal convection. Phys. Rev. Fluids **3**, 053501 (2018)
43. Lüthje, M., Feltham, D.L., Taylor, P.D., Worster, M.G.: Modeling the summertime evolution of sea–ice melt ponds. J. Geophys. Res. **111**, C02001 (2006)
44. Lüthje, M., Pedersen, L.T., Reeh, N., Greuell, W.: Modelling the evolution of supraglacial lakes on the West Greenland ice-sheet margin. J. Glaciol. **52**(179), 608–618 (2006)
45. Skyllingstad, E.D., Paulson, C.A., Perovich, D.K.: Simulation of melt pond evolution on level ice. J. Geophys. Res. **114**, C12019 (2009)
46. Scott, F., Feltham, D.L.: A model of the three-dimensional evolution of Arctic melt ponds on first-year and multiyear sea ice. J. Geophys. Res. **115**, C12064 (2010)
47. Thorndike, A.S., Rothrock, D.A., Maykut, G.A., Colony, R.: The thickness distribution of sea ice. J. Geophys. Res. **80**(33), 4501–4513 (1975)
48. Hunke, E.C., Lipscomb, W.H.: CICE: The Los Alamos Sea Ice Model. Documentation and software user's manual version 4.0. Tech. Rep. LA-CC-06-012. T-3 Fluid Dyn. Group, Los Alamos Natl. Lab., Los Alamos, NM (2008)
49. Vancoppenolle, M., Bouillon, S., Fichefet, T., Goosse, H., Lecomte, O., Morales Maqueda, M.A., Madec, G.: The Louvain-la-Neuve sea ice model. Notes du pole de modélisation, Institut Pierre-Simon Laplace (IPSL), Paris (2012)
50. Flocco, D., Feltham, D.L.: A continuum model of melt pond evolution on Arctic sea ice. J. Geophys. Res. **112**, C08016 (2007)
51. Flocco, D., Feltham, D.L., Turner, A.K.: Incorporation of a physically based melt pond scheme into the sea ice component of a climate model. J. Geophys. Res. **114**, C08012 (2010)
52. Ahlers, G., Grossmann, S., Lohse, D.: Heat transfer and large scale dynamics in turbulent Rayleigh-Bénard convection. Rev. Mod. Phys. **81**, 503 (2009)
53. Kim, J.-H., Moon, W., Wells, A.J., Wilkinson, J.P., Langton, T., Hwang, B., Granskog, M.A., Rees Jones, D.W.: Salinity control of thermal evolution of late summer melt ponds on Arctic sea ice. Geophys. Res. Lett. **45** (2018). https://doi.org/10.1029/2018GL078077
54. Grossmann, S., Lohse, D.: Scaling in thermal convection: a unifying theory. J. Fluid Mech. **407**, 27–56 (2000)
55. Malkus, M.V.R.: The heat transport and spectrum of thermal turbulence. Proc. R. Soc. Lond. A **225**, 196 (1954)
56. Rallabandi, B., Zheng, Z., Winton, M., Stone, H.A.: Formation of sea ice bridges in narrow straits in response to wind and water stresses. J. Geophys. Res. Oceans **122**(7), 5588–5610 (2017)
57. Domaradzki, J.A., Metcalfe, R.W.: Direct numerical simulations of the effects of shear on turbulent Rayleigh-Bénard convection. J. Fluid Mech. **193**, 499 (1988)
58. Scagliarini, A., Gylfason A., Toschi, F.: Heat-flux scaling in turbulent Rayleigh-Bénard convection with an imposed longitudinal wind. Phys. Rev. E **89**, 043012 (2014)
59. Prandtl, L.: Bericht über die Entstehung der Turbulenz. Z. Angew. Math. Mech. **5**, 136–139 (1925)
60. Tsamados, M., Feltham, D.L., Petty, A.A., Schröder D., Flocco, D.: Processes controlling surface, bottom and lateral melt of Arctic sea ice in a state of the art sea ice model. Philos. Trans. R. Soc. A **373**, 20140167 (2015)
61. Eicken, H., Krouse, H.R., Kadko, D., Perovich, D.K.: Tracer studies of pathways and rates of meltwater transport through Arctic summer sea ice. J. Geophys. Res. **107**(C10), 8046 (2002)
62. Landau, L.D., Lifshitz, E.M.: Fluid Mechanics, 2nd edn. Pergamon Press, Oxford (1987)

63. Marche, F.: Derivation of a new two-dimensional viscous shallow water model with varying topography, bottom friction and capillary effects. Eur. J. Mech. B Fluids **26**, 49–63 (2007)
64. Oron, A., Davis, S.H., Bankoff, S.G.: Long-scale evolution of thin liquid films. Rev. Mod. Phys. **69**(3), 931–980 (1997)
65. Hvidegaard, S.M., Forsberg, R.: Sea-ice thickness from airborne laser altimetry over the Arctic Ocean north of Greenland. Geophys. Res. Lett. **29**(20), 1952 (2002)
66. Hoshen, J., Kopelman, R.: Percolation and cluster distribution. I. Cluster multiple labeling technique and critical concentration algorithm. Phys. Rev. B. **14**, 3438–3445 (1976)
67. Moritz, R.E., Curry, J.A., Thorndike, A.S., Untersteiner, N.: SHEBA a Research Program on the Surface Heat Budget of the Arctic Ocean. Rep. 3, 34 pp., Arctic Syst. Sci.: Ocean-Atmos.-Ice Interact. (1993)
68. Moritz, R.E., Perovich, D.K. (eds.): Surface Heat Budget of the Arctic Ocean, Science Plan, ARCSS/OAII. Rep. 5, 64 pp., Univ. of Wash., Seattle (1996)
69. Perovich, D.K., Grenfell, T.C., Light, B., Elder, B.C., Harbeck, J., Polashenski, C., Tucker III, W.B., Stelmach, C.: Transpolar observations of the morphological properties of Arctic sea ice. J. Geophys. Res. **114**, C00A04 (2009)
70. Ma, Y.-P., Sudakov, I., Strong, C., Golden, K.M.: Ising model for melt ponds on Arctic sea ice (2014). arXiv:1408.2487
71. Popović, P., Cael, B.B., Silber, M., Abbot, D.S.: Simple rules govern the patterns of arctic sea ice melt ponds. Phys. Rev. Lett. **120**, 148701 (2018)
72. Hohengger, C., Alali, B., Steffen, K.R., Perovich, D.K., Golden, K.M.: Transition in the fractal geometry of Arctic melt ponds. Cryosphere **6**, 1157–1162 (2012)
73. Mandelbrot, B.B.: The Fractal Geometry of Nature. Freeman, New York (1982)
74. Cheng, Q.: The perimeter-area fractal model and its application to geology. Math. Geol. **27**, 69–84 (1995)
75. Grassberger, P.: Generalized dimensions of strange attractors. Phys. Lett. **97A**(6), 227–230 (1993)

Part IV
Theme: Ecology

Carbon Dioxide Time Series Analysis: A New Methodological Approach for Event Screening Categorisation

Stefano Bianchi, Wolfango Plastino, Alcide Giorgio di Sarra,
Salvatore Piacentino, and Damiano Sferlazzo

Abstract A new method for time series analysis allowing the description of
background evolution and outliers was developed. This approach was tested on
CO_2 weekly measurements made at Mauna Loa, Hawaii, and Lampedusa, Italy,
for a period (1992–2014) longer than the 11-year solar cycle. After the time
series was detrended, the Generalised Lomb-Scargle periodogram was computed
for frequency domain analysis. All frequencies corresponding to values in the
spectrum higher than a threshold were filtered out, and the time series was reduced
to residuals. Residuals were analysed principally focusing on persistency, inspected
via detrended fluctuation analysis. The analysis allows to highlight similarities and
differences between the two stations. Annual and semi-annual periods are present
in both the time series, with significantly larger amplitudes at Lampedusa than at
Mauna Loa, where however they explain about 83% of the variability, compared to
about 62% at Lampedusa. Remarkably, a different Hurst exponent was found, with
a value corresponding to pink noise for Mauna Loa, and a smaller value (about 0.80)
for Lampedusa. This is attributed to the different characteristics of the two stations.

Keywords CO_2 · Time series analysis · Hurst exponent

1 Introduction

The observed increase of atmospheric CO_2 is the result of various processes which
include natural variability in emission and absorption, anthropic emissions, atmospheric transport and photochemical processes, and complex interactions among

S. Bianchi (✉) · W. Plastino
Department of Mathematics and Physics, Roma Tre University, Rome, Italy
e-mail: stefano.bianchi@uniroma3.it; wolfango.plastino@uniroma3.it

A. G. di Sarra · S. Piacentino · D. Sferlazzo
Laboratory for Observations and Analyses of the Earth and Climate, ENEA, Rome, Italy
e-mail: alcide.disarra@enea.it; salvatore.piacentino@enea.it; damiano.sferlazzo@enea.it

© Springer Nature Switzerland AG 2020
P. Cannarsa et al. (eds.), *Mathematical Approach to Climate Change
and Its Impacts*, Springer INdAM Series 38,
https://doi.org/10.1007/978-3-030-38669-6_7

vegetation, the atmosphere, oceans, and, over very long time scales, geological structures. A relatively large number of sites has been setup starting from 1950s for the monitoring of atmospheric CO_2 and the investigation of these complex processes and quantifying the human impact on climate. Several methods have been implemented to analyse these time series of CO_2 surface observations [1, 9, 12, 13]. The first three principally focus on the variability of the main periodicities and on trend estimation via filtering methods, while the last one studies long-range correlations of Mauna Loa residuals, after removing trend and periodicities by means of a polynomial and sinusoidal fit. Moreover, these methods require data interpolation where missing data are present. In this study we apply a recently developed methodology for the analysis of irregular/discontinuous time series [2] to long-term CO_2 records at two sites with markedly different characteristics. The analysis gives information on the time series in terms of data-driven trend, periodicities (predominant and smaller ones), and correlation properties of the fluctuations. A complete description of the time series properties is therefore given, making the analysis a good choice when a comparison among different sites is done.

The analysis of the same variable in different locations is aimed at highlight similarities, possibly linked to global processes, and differences related with regional and local phenomena. The first site is Lampedusa ($35.52°$ N, $12.63°$ E), a small island in the central Mediterranean sea. The measurement site is at an altitude of 45 m a.s.l., and is operative since 1992. The second station is Mauna Loa ($19.54°$ N, $155.58°$ W), where CO_2 monitoring started in the 1950's at the remarkable height of 3397 m a.s.l. Its remote location, height, and minimal influences from vegetation and human activity are ideal for CO_2 monitoring and analyses as far as climate change is concerned.

The paper is organised as follows. Section 2 explains the scheme of the methodological approach and the mathematical tools employed for the analysis. Section 3 presents the results for the comparison between Lampedusa and Mauna Loa stations.

2   Methods

The methodology for the time series analysis is here summarised. First, the trend of the time series is estimated by means of Empirical Mode Decomposition (EMD) [4]. EMD is a data-driven adaptive method. Local maxima and minima in the signal are identified and subsequently interpolated with a cubic spline, defining an upper and a lower envelope for the signal. The mean envelope is subtracted from the signal and the procedure is repeated until the number of extrema is equal to the number of zeros plus or minus one. The so-obtained Intrinsic Mode Function (IMF) is then subtracted from the original time series and the whole procedure is reapplied to the new signal (original time series minus IMF). The process stops when a monotonic function is obtained and the time series X_t can be decomposed as the sum of the IMFs plus the trend T_t (the monotonic function).

Then, following Bianchi et al. [2], the Generalised Lomb-Scargle (GLS) periodogram [8, 14],

$$P(\omega) = \frac{1}{2\sigma^2} \left\{ \frac{N[\sum_j (X_j - \langle X \rangle) \cos(\omega(t_j - \tau))]^2}{N \sum_j \cos^2(\omega(t_j - \tau)) - [\sum_j \cos(\omega(t_j - \tau))]^2} + \right.$$
$$\left. + \frac{N[\sum_j (X_j - \langle X \rangle) \sin(\omega(t_j - \tau))]^2}{N \sum_j \sin^2(\omega(t_j - \tau)) - [\sum_j \sin(\omega(t_j - \tau))]^2} \right\} \quad (1)$$

can be evaluated in order to extract periodic components and quantify their percentage weight with respect to the total time series. X_j are the CO_2 weekly data at time t_j, N is the length of the time series, $\omega = 2\pi \nu$, with ν equal to the frequency, σ^2 the variance of the data, $\langle X \rangle$ the mean of the data. τ is given by [11, 14]

$$\tau = \frac{1}{2\omega} \arctan \left\{ \frac{N \sum_j \sin(2\omega t_j) - 2 \sum_j \cos(\omega t_j) \sum_j \sin(\omega t_j)}{N \sum_j \cos(2\omega t_j) - [\sum_j \cos(\omega t_j)]^2 + [\sum_j \sin(\omega t_j)]^2} \right\} \quad (2)$$

The GLS periodogram can handle cases of missing data without any interpolation, and it is also associated with a threshold that allows to discriminate peaks with high probability to be associated to a periodic signal. Once peaks corresponding to periodicities have been removed, residual time series can be investigated. First, it is arranged in a histogram, and then normalised to zero mean and unit variance. The values that exceed a certain number of standard deviations are considered as outliers, i.e. anomalous values.

Finally, the persistence of the residual time series is inspected via Detrended Fluctuation Analysis (DFA) [10]. The residual time series is first integrated, obtaining a new time series Y_t, and then divided in time intervals of length n. The data in every interval are fitted with a least squares line Y_t^{fit} and the root mean square fluctuation in each interval of length n is computed. Then, all the root mean square fluctuations are averaged, and the whole procedure is repeated for different sizes n of the time interval, obtaining

$$F(n) = \left[\frac{1}{\lfloor N/n \rfloor} \sum_{s=1}^{\lfloor N/n \rfloor} \left[\frac{1}{n} \sum_{t=1}^{n} (Y_t - Y_t^{fit})^2 \right] \right]^{1/2} \sim n^H \quad (3)$$

H being the Hurst exponent. Values of H between 0.0 and 0.5 indicate anti-correlation, $H = 0.5$ corresponds to white noise, $0.5 < H < 1.0$ indicates long range correlations, $H = 1.0$ is the equivalent of pink noise (i.e. power spectrum going as the inverse of the frequency), $1.0 < H < 1.5$ indicates stronger long range correlations, and $H = 1.5$ corresponds to Brownian noise. Also, a local Hurst exponent H_t can be defined [5] as follows. Generalising Eq. (3) to order q,

$$F_q(n) = \left[\frac{1}{\lfloor N/n \rfloor} \sum_{s=1}^{\lfloor N/n \rfloor} \left[\frac{1}{n} \sum_{t=1}^{n} (Y_t - Y_t^{fit})^2 \right]^{q/2} \right]^{1/q} \sim n^{h(q)} \qquad (4)$$

where $s = 1, \ldots, \lfloor N/n \rfloor$ is the number of intervals with length n. $h(2) = H$ is the Hurst exponent. Given N the length of the time series, $F_q(N)$ is common to all the orders q. Taking a small window that runs over the total length of the time series, root mean square fluctuations can be computed in every window in order to obtain the values of the fluctuations as a function of time. Then, these values and the value of $F_q(N)$ can be connected with a straight line to obtain a local Hurst exponent H_t as the slope of the line, i.e. the Hurst exponent for a single time window.

3   Results and Discussion

The CO_2 time series used in this study range from May 1992 to November 2014 for both Lampedusa and Mauna Loa. The dataset is constituted by weekly values and both time series contain some missing data. Ambient air is sampled in flasks at the two sites, and subsequently analysed in the lab [3, 7]. Even though CO_2 values range between 350 and 400 parts per million (ppm) for both the time series, the Lampedusa data appear to be more noisy, with higher variations than at Mauna Loa. This is reflected in the trend T_t, as shown in the upper part of Fig. 1. The trend for Lampedusa is less steady, due to years with a smaller amplitude. Once the trend is removed, the difference in amplitude is more evident, as shown in the lower part of Fig. 1. Here the amplitude of the Lampedusa time series is about 20 ppm, while Mauna Loa time series only has half the amplitude.

An opposite result is found in the frequency domain, where the Mauna Loa time series gives higher contribution in terms of periodicities with respect to the

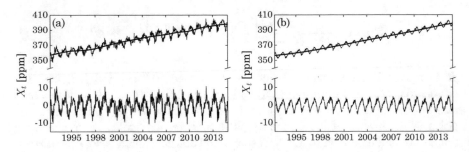

Fig. 1 Upper part: Trend estimation via empirical mode decomposition for CO_2 time series measured in Lampedusa (**a**) and Mauna Loa (**b**). The trend T_t is the solid line superimposed on the time series. Lower part: Detrended time series for Lampedusa (**a**) and Mauna Loa (**b**)

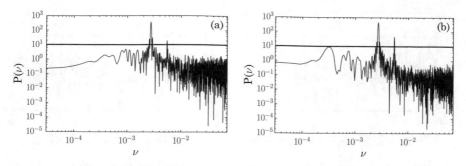

Fig. 2 LS periodogram for Lampedusa (**a**) and Mauna Loa (**b**) time series. Peaks over the threshold (horizontal solid line) correspond to the annual and semi-annual cycles

Lampedusa time series. The GLS periodogram is shown in Fig. 2. Two peaks over the threshold are present in both spectra and correspond to the annual and semi-annual cycles. The annual peak is prominent in both periodograms, and explains 76.5% of the variability of the Mauna Loa time series, and 59.4% of the variability of the Lampedusa time series. Thus, even though the annual cycle of the Lampedusa time series has a bigger amplitude in the time domain, it contributes less in the frequency domain compared to the one of Mauna Loa. Similarly, the semi-annual cycle, contributes by 6.6% to the periodogram of Mauna Loa, and by 2.5% to the periodogram of Lampedusa. If the annual and semi-annual cycles are removed, the GLS periodogram can be reapplied until no peaks above threshold are found. For Lampedusa, the other peaks correspond to periodicities of $\sim$3.5, $\sim$3.0, $\sim$1.8 years, all contributing by less than 1% to the periodogram. As emphasised by Chamard et al. [3] and Artuso et al. [1], a correlation exists between the CO_2 growth rate at Lampedusa and global temperature and El Niño-Southern Oscillation, and these correlations involve similarities in the 3 and 3.5 year periodicities. For Mauna Loa instead, periodicities are $\sim$8.2, $\sim$4.7, $\sim$3.8, $\sim$3.0, $\sim$2.4 years, all contributing by less than 1% to the periodogram, except for the first one which has a percentage weight of 1.7%. Kane and de Paula [6] analysed Mauna Loa monthly CO_2 observations over the period 1959–1992 and found highly significant periodicities of 14, 8, 5, and 3.4 years, as well as at 2.6 years, which appears to be connected with the Southern Oscillation. They also found a periodicity on the annual amplitude with seems to be related with the Quasi Biennial Oscillation. The frequency at about 3.4 years was found also in the global temperature time series. The similarities in some of the frequencies indicate that large scale/global processes, such as variations in the global temperature and the Southern Oscillation, are related with the CO_2 evolution at least on the hemispheric scale. Some processes, like the quasi biennial oscillation, appear to specifically pertain the tropical regions and to have a negligible influence on CO_2 at mid latitudes.

Peaks corresponding to the annual and semi-annual cycles are removed, and residual time series are normalised to zero mean and unit variance, since both time series passed the test against a normal distribution. Normalised residual time series

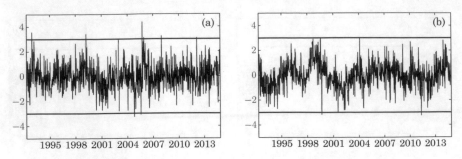

Fig. 3 Normalised residual for Lampedusa (**a**) and Mauna Loa (**b**) time series. The two horizontal lines represent the $\pm 3\sigma$ thresholds

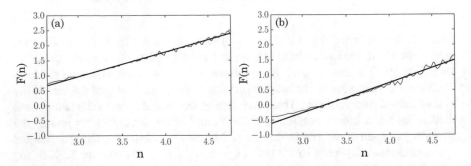

Fig. 4 DFA for Lampedusa (**a**) and Mauna Loa (**b**) time series. The best fit is represented by the solid straight line, whose slope is $H = 0.811 \pm 0.008$ at Lampedusa and $H = 0.994 \pm 0.014$ at Mauna Loa

are shown in Fig. 3. As expected, the Mauna Loa residuals time series is essentially devoid of outliers, and is less noisy than at Lampedusa. The Lampedusa residual time series is instead noisier and an outlier above $+4\sigma$ occurs on 16 September 2005. Correlative measurements and analyses with backward airmass trajectories suggest that local gaseous sources have influenced the sampled air, which contains relatively high CH_4. The chemical composition of the aerosol sampled on the same day suggests that the dominant sources of particles are from the marine and the biogenic sector. Standard deviations used for normalisation are 2.23 and 0.78 ppm for Lampedusa and Mauna Loa, respectively, confirming the larger fluctuations in CO_2 records in the Italian station. This behaviour is well explained by the influence from regions with contrasting characteristics with an uneven distribution of sources (mainly anthropic emissions, in particular from central and eastern Europe, fires in summer, ship traffic, and desert areas in the south) and sinks (mainly vegetation in Europe, and desert in Africa), the latitudinal CO_2 gradient, and modulation by synoptic disturbances [1].

Finally, persistence is inspected via detrended fluctuation analysis (DFA), both globally and locally. A minimum window of about 3 months has been chosen for the analysis. Results of DFA are shown in Fig. 4, where once again fluctuations

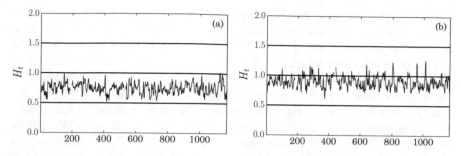

Fig. 5 Local Hurst exponent H_t for Lampedusa (**a**) and Mauna Loa (**b**) time series. The three horizontal lines represent the values for red noise (top), pink noise (middle), and white noise (bottom)

at Lampedusa are larger than those at Mauna Loa. The Hurst exponent is $H = 0.994 \pm 0.014$ at Mauna Loa, i.e. residuals are in the pink noise regime, and $H = 0.811 \pm 0.008$ at Lampedusa. Thus, even though both residual time series display long-range correlations, Mauna Loa exhibits stronger correlations than Lampedusa. The difference in the Hurst exponent, i.e. the difference in the scaling properties of the residual time series, is associated with the factors that can influence the CO_2 concentrations in the two sites. Mauna Loa station is placed at a very high altitude, and has minimal influences coming from vegetation and human activity. Similar results on the long-range correlation were found at Mauna Loa by Varotsos et al. [13]. Conversely, this long-range correlation might be masked by the higher shorter-term variability at Lampedusa.

Then, the local Hurst exponent H_t is evaluated for the 3 months time window. Results are displayed in Fig. 5. The value of the local Hurst exponent fluctuates in time, with low values corresponding to periods with great CO_2 variations, and high values corresponding to periods with small variations. The mean value of H_t is slightly smaller than the Hurst exponent H, as can be seen in Fig. 4, where for a time window of 3 months $F(n)$ deviates from the straight line fit.

4 Conclusions

A new method for time series analysis has been applied to weekly CO_2 data from Mauna Loa and Lampedusa over the period 1992–2014. The method can be applied to not-evenly spaced or discontinuous datasets and allows to characterise the data evolution and variability on different time scales. The analysis performed on two sites with markedly different characteristics allows to identify some aspects of the CO_2 variability, the influencing processes, and the site-specific regimes. Annual and semi-annual cycles play an important role at the two sites, and explain about 83% of the variability at Mauna Loa, and about 62% at Lampedusa. The annual amplitude is about two-times larger at Lampedusa than at Mauna Loa, as well as the short term variability. The analysis allows to identify main frequencies present in the datasets. Frequencies of global-hemispheric phenomena (such as the variation of the global temperature, and the Southern Oscillation) are found in both datasets. Other specific periodicities are characteristics of each site (i.e., the quasi biennial oscillation at Mauna Loa, and a shorter term variability at Lampedusa). Normalised CO_2 residual time series are obtained after removal of main periodic signals, for outliers identification. Residuals are significantly smaller at Mauna Loa than at Lampedusa, with essentially no outliers. The detrended fluctuation analysis was applied to the residual, with the aim of determining variability regime. The determination of the Hurst exponent suggests that a long-term correlation, possibly non existent or masked by the high variability at Lampedusa, is present in the Mauna Loa data. These results, which in part confirm those obtained in previous studies separately for Mauna Loa and Lampedusa, provide useful information for the characterisation of the two sites and the understanding of processes which influence the CO_2 on different spatial and temporals scales. This analysis also suggests that the proposed methodology is a useful tool for the interpretation of long-term CO_2 data.

Acknowledgements Thanks are due to the National Oceanographic and Atmospheric Administration/Earth System Research Laboratory for providing weekly CO_2 data at Mauna Loa. Data have been downloaded from ftp://aftp.cmdl.noaa.gov/data/trace_gases/co2. Measurements at Lampedusa have been partly supported by the Italian Ministry for University and Research through the NextData Project, and by the European Union through Projects CarboEurope, IMECC, and GHG-Europe. Contributions by Francesco Monteleone are gratefully acknowledged.

References

1. Artuso, F., Chamard, P., Piacentino, S., Sferlazzo, D.M., De Silvestri, L., di Sarra, A., Meloni, D., Monteleone, F.: Influence of transport and trends in atmospheric CO_2 at Lampedusa. Atmos. Environ. **43**, 3044–3051 (2009)
2. Bianchi, S., Longo, A., Plastino, W., Povinec, P.: Evaluation of ^7Be and ^{133}Xe atmospheric radioactivity time series measured at four CTBTO radionuclide stations. Appl. Radiat. Isot. **132**, 24–28 (2018)
3. Chamard, P., Thiery, F., di Sarra, A., Ciattaglia, L., De Silvestri, L., Grigioni, P., Monteleone, F., Piacentino, S.: Interannual variability of atmospheric CO_2 in the Mediterranean: measurements at the island of Lampedusa. Tellus **55B**, 83–93 (2003)
4. Huang, N.E., Shen, Z., Long, S.R., Wu, M.C., et al.: The empirical mode decomposition and the Hilbert spectrum for nonlinear and non-stationary time series analysis. Proc. R. Soc. Lond. A Math. Phys. Eng. Sci. **454** 903–995 (1998)
5. Ihlen, E.A.: Introduction to multifractal detrended fluctuation analysis in Matlab. Fractal Anal. **3**, 97 (2012)
6. Kane, R.P., de Paula, E.R.: Atmospheric CO_2 changes at Mauna Loa, Hawaii. J. Atmos. Terr. Phys. **58**, 1673–1681 (1996)
7. Komhyr, W.D., Gammon, R.H., Harris, T.B., Waterman, L.S., Conway, T.J., Taylor, W.R., Thoning, K. W.: Global atmospheric CO_2 distribution and variations from 1968–1982 NOAA/GMCC CO_2 flask sample data. J. Geophys. Res. **90**, 5567–5596 (1985)
8. Lomb, N.R.: Least-squares frequency analysis of unequally spaced data. Astrophys. Space Sci. **39**(2), 447–462 (1976)

9. Nakazawa, T., Ishizawa, M., Higuchi, K., Trivett, N.B.A.: Two curve fitting methods applied to CO_2 flask data. Environmetrics **8**, 197–218 (1997)

10. Peng, C.K., Buldyrev, S.V., Goldberger, A.L., Havlin, S., Simons, M., Stanley, H.E.: Finite-size effects on long-range correlations: implications for analyzing DNA sequences. Phys. Rev. E **47**(5), 3730 (1993)

11. Press, W.H., Rybicki, G.B.: Fast algorithm for spectral analysis of unevenly sampled data. Astrophys. J. **338**, 277–280 (1989)

12. Thoning, K.W., Tans, P.P., Komhyr, W.D.: Atmospheric carbon dioxide at Mauna Loa Observatory, 2. Analysis of the NOAA/GMCC data, 1974–1985. J. Geophys. Res. **94**, 8549–8565 (1989)

13. Varotsos, C., Assimakopoulos, M.N., Efstathiou, M.: Technical note: long-term memory effect in the atmospheric CO_2 concentration at Mauna Loa. Atmos. Chem. Phys. **7**, 629–634 (2007)

14. Zeichmeister, M., Kürster, M.: The generalised Lomb-Scargle periodogram: a new formalism for the floating-mean and Keplerian periodograms. Astron. Astrophys. **496**(2), 577–584 (2009)

Mathematical Tools for Controlling Invasive Species in Protected Areas

Carmela Marangi, Francesca Casella, Fasma Diele, Deborah Lacitignola,
Angela Martiradonna, Antonello Provenzale, and Stefania Ragni

Abstract A challenging task in the management of Protected Areas is to control
the spread of invasive species, either floristic or faunistic, and the preservation of
indigenous endangered species, typically competing for the use of resources in a
fragmented habitat. In this paper, we present some mathematical tools that have been
recently applied to contain the worrying diffusion of wolf-wild boars in a Southern
Italy Protected Area belonging to the Natura 2000 network. They aim to solve the
problem according to three different and in some sense complementary approaches:
(i) the qualitative one, based on the use of dynamical systems and bifurcation
theory; (ii) the Z-control, an error-based neural dynamic approach; (iii) the optimal
control theory. In the case of the wild-boars, the obtained results are illustrated
and discussed. To refine the optimal control strategies, a further development is to
take into account the spatio-temporal features of the invasive species over large and
irregular environments. This approach can be successfully applied, with an optimal
allocation of resources, to control an invasive alien species infesting the Alta Murgia
National Park: *Ailanthus altissima*. This species is one of the most invasive species

C. Marangi (✉) · F. Diele · A. Martiradonna
Istituto per le Applicazioni del Calcolo 'M. Picone', CNR, Bari, Italy
e-mail: c.marangi@ba.iac.cnr.it; f.diele@ba.iac.cnr.it; a.martiradonna@ba.iac.cnr.it

F. Casella
Istituto di Scienze delle Produzioni Alimentari, CNR, Bari, Italy
e-mail: francesca.casella@ispa.cnr.it

D. Lacitignola
Dipartimento di Ingegneria Elettrica e dell'Informazione, Università di Cassino e del Lazio
meridionale, Cassino, Italy
e-mail: d.lacitignola@unicas.it

A. Provenzale
Istituto di Geoscienze e Georisorse, CNR, Pisa, Italy
e-mail: antonello.provenzale@igg.cnr.it

S. Ragni
Department of Economics and Management, University of Ferrara, Ferrara, Italy
e-mail: stefania.ragni@unife.it

© Springer Nature Switzerland AG 2020
P. Cannarsa et al. (eds.), *Mathematical Approach to Climate Change
and Its Impacts*, Springer INdAM Series 38,
https://doi.org/10.1007/978-3-030-38669-6_8

in Europe and its eradication and control is the object of research projects and biodiversity conservation actions in both protected and urban areas [11]. We lastly present, as a further example, the effects of the introduction of the brook trout, an alien salmonid from North America, in naturally fishless lakes of the Gran Paradiso National Park, study site of an on-going H2020 project (ECOPOTENTIAL).

Keywords Invasive species · Dynamical systems · Optimal control

1 Introduction

In many protected areas, according to their regulation framework, the exploitation of ecosystem services produces a disturbance factor which demands for counteractions in order to guarantee the equilibria of the areas relevant ecological systems. A clear example is the introduction of alien species for leisure hunting/fishing purposes that have become invasive. Invasive alien species are species that have been transported by humans outside of their natural range, have become established, and are causing ecological damage to the recipient ecosystem [16]. Invasive species cost to European Union at least 12 billion of euros per year [24], including costs for damages in key economic sectors, such as agriculture, fisheries, aquaculture, forestry and health. Invasive species cause not only economic damages but are also a major cause of global biodiversity loss [47]. Controlling or even eliminating invasive species can have enormous benefits [53] but, unfortunately, it is a costly endeavor which requires careful planning to ensure cost-effectiveness [54]. The managing authorities of the protected areas have to face the challenge to make the fruition of the natural capital sustainable, by balancing the immediate benefits it can provide with the conservation targets to make the natural capital still available for future generations. That requires a number of actions to be put in place to obtain, preserve and enforce equilibrium states for ecosystems. The present paper mainly focuses on the Alta Murgia National Park in Italy, a node of the Natura 2000 network and the spread of wild boars (*Sus scrofa*) is considered as case study. Moreover, in the same protected area we will present some theoretical results about a model for the control of tree of heaven (*Ailanthus altissima*). Finally, a more general model for the control of the brook trout (*Salvelinus fontinalis*) in Alpine Lakes of Gran Paradiso National Park is also illustrated. As we will show, from the mathematical point of view, the problem of the spreading of an invasive species can be attacked through three approaches, different for philosophies and methodologies: (i) dynamical systems theory; (ii) the so-called *Z-control*; (iii) the optimal control theory.

The first approach—based on the so called *qualitative analysis*—deals with mathematical models within the framework of population dynamics. Since such models can be commonly described as nonlinear systems of differential equations [28, 55], dynamical systems theory has become an important tool for investigating fundamental questions related to their long-term phenomenology [46, 49, 57]. Qualitative analysis is essentially devoted to prove the existence and stability of the system attractors as well as to search for those mechanisms responsible of qualitative changes in the dynamics by performing bifurcation analysis. The aim is to obtain analytical conditions, expressed in terms of the system parameters, that determine long-term behavior of the system and—once such conditions are obtained—act on the most suitable parameter to *control* the dynamics. For example, in the most simple case, the control of invasive species can be put within the framework of harvested populations that have commonly been studied with using the control parameter as the bifurcation parameter, e.g. [34] and references therein. In this case, the dynamical system approach, using qualitative analysis, can provide qualitative indications on the mechanisms that are responsible of specific system behaviors, hence suggesting possible strategies to avoid some undesirable situations.

The second approach, the *Z-type dynamic method* [29, 35, 58, 59], is an error-based dynamic approach in which the key element is the design formula which ensures that each element of the error function converges to zero exponentially. Since the Z-type control laws have exponential convergence performances, the convergence time can be estimated and populations will converge to the desired state in a fast or predefined rate. This feature enables the Z-type dynamic method to be particularly efficient for specific time limited applications.

The third approach makes use of *dynamic optimization methods*, whose aim is maximizing the returns and/or minimizing the costs involved in the management of social, economic or ecological processes. Broadly, these methods can be deterministic, such as optimal control theory [37, 38], or stochastic, such as stochastic dynamic programming [42]. In this paper we focus on deterministic optimal control theory. A large amount of the optimal control literature focuses in fact on the control of invasive species. For example, in [50] the optimal strategies to control scotch broom in an Australian national park was investigated; in [14] it was considered the control of Californian thistle in New Zealand agriculture; in [10] the optimal management options for brown tree snakes in Hawaii was considered; in [13] it was studied the control of blackberry in Australia; in [4] the case study of Feral cats (*Felis catus*) in Australia was analyzed.

Optimal control deals in fact with the problem of finding a control law for a given system such that a certain optimality criterion is achieved. The ingredients of the control problem are: (i) the objective function, which is a function of the state and control variables; and (ii) a set of constraints—in our case of differential equations—that the state variables must satisfy. An optimal control path is the path of the control variable that minimizes or maximizes the objective function. Once the optimal path is found, paths for the state variables follow. Optimal control problems are usually nonlinear and, since it is usually impossible to find analytic solutions, numerical methods are usually implemented. In practice: (i) by using Pontryagin's maximum principle [51], one obtains the first-order optimality conditions which result in a boundary value problem for an Hamiltonian dynamical system—the state-costate model—involving the state variable and the *adjoint* auxiliary variable; and (ii) this boundary value problem for the state and the adjoint variable is numerically solved by using the appropriate boundary (or transversality) conditions, and the resulting solution is readily verified to be an extremal trajectory. Indirect methods for solving optimal control problems are based on Pontryagin's maximum principle and solve the associated Hamiltonian boundary value problem that constitutes the first-order necessary conditions for optimality. Structure-preserving algorithms as symplectic integrators have been proposed in literature for the numerical solution of the Hamiltonian system [31]. The role of symplectic integrators in optimal control has been investigated in [15] and, more recently, in [52]. Partitioned Runge-Kutta discretizations based on symplectic pairs, for which the direct and indirect approaches are equivalent have been proposed in [8, 18, 19, 30].

In the following sections we briefly recall some ecological and management issues related to the specific invasive species considered in this paper, i.e. wild boars (*Sus scrofa*), tree of heaven (*Ailanthus altissima*) and brook trout (*Salvelinus fontinalis*). Although in most of the examples here shown we refer just to the wild boars as invasive species, that gives a better view of the different problems to be faced with the different approaches here proposed. We then review the different mathematical control approaches—as developed in details in [5, 35, 36, 44]—when applied to the wild boars population spreading in the Alta Murgia National Park. In the last part of the paper, we illustrate and discuss a spatial model introduced in [6] that will be specialized for the control problem of the tree of heaven in the Alta Murgia National Park and of the brook trout in the Alpine Lakes of Gran Paradiso National Park.

2 Invasive Species: Ecological and Management Issues

2.1 Wild Boars (Sus scrofa)

The establishment of management policies for protected areas has to face the issue of crop damages originating from wildlife. These damages have the negative effect of increasing the intolerance toward the wildlife and of diverting a substantial part of the usually scarce resources allocated to nature conservation toward compensation actions [26]. In several countries Wild boars are deemed responsible for the largest amount of damages by wildlife. However, their heavy impact on agriculture practices may be partially compensated by their positive action on the native biotic communities of the ecosystem they are embedded in. To reduce the size of the wild boar population and its ensuing impact on agriculture, while at the same time retaining the advantages of their presence in protected areas, culling activities need to be carefully planned by the managing authorities. A relevant issue in a planning task is that wild boars play a significant role in the trophic chain of another species, the wolf, which is currently at risk of extinction [45]. Wild boars represent in fact the favorite prey of wolves, especially in areas where other preys such as wild ungulates, are absent. Wolves are considered keystone species, as their impact on their embedding ecosystems is disproportionally large with respect to their abundance, since their extinction can cause a cascade of secondary extinctions. The legal status of wolves in the European Union countries is directly specified in the Habitats Directive (92/43/EEC)2 which is also responsible for the establishment of the Natura 2000 network of protected areas. The Directive requires strict protection, prohibiting any destruction or damage to the wolf population [33]. This is reflected, for example, in the plan for the management of wild boars, in the Alta Murgia National Park (see Delibera n. 21/2012 del 18/12/2012 Presidenza Parco Nazionale dell'Alta Murgia), a protected area where the ecological and management problem to be faced is how to control the wild boars population ensuring at the same time the preservation of the wolves.

2.2   Tree of Heaven (Ailanthus altissima)

In Apulia Region (South Italy) *Ailanthus altissima* (Miller) Swingle is the most widespread and harmful invasive alien plant species in non-crop areas. It is native from Asia and was introduced into Europe in the eighteenth century for the breeding of the silkworm moth *Philosamia cynthia* [12]. Due to its rapid growth, robust root system, adaptability to any type of soil and water regime and lack of natural enemies, in the twentieth century it was broadly used in Italy for ornamental purposes and for the consolidation of escarpments. It spreads spontaneously colonizing new areas at considerable distance from the parent plant through a high number of winged samaras, transported by wind and water. It also propagate by root suckers, forming highly dense stands out-competing other species and reducing their growth. Invasive alien species such as ailanthus are considered one of the most significant cause of global biodiversity loss. A. *altissima* causes serious direct and indirect ecological, economic, functional and aesthetic damages in non-crop areas [11]. Take a census and mapping the invasive species is critical to quantify the management costs and to carry on a control program. In the Alta Murgia National Park (Apulia Region, South Italy) within the Natura 2000 Network, tree of heaven is highly present and threatening the fragile grasslands ecosystem. In that area the LIFE Programme, the EU's financial instrument for environmental and nature conservation, funded the LIFE Alta Murgia project (LIFE12 BIO/IT/000213) which has, as main objective, the eradication of the invasive exotic tree species A. *altissima* from the Alta Murgia National Park using innovative and environmentally friendly techniques. The control problem here is the one of eradication/reduction of the invasive species with an optimization of the costs of the control program. In many cases that has to be done taking into account a budget constraint.

2.3   Brook Trout (Salvelinus fontinalis)

Several protected areas, before their establishment as such, have been hosting hunting/fishery activities for leisure purposes. To sustain these activities, a common practice was the massive introduction of preys, sometimes consisting in alien species, with a high degree of adaptability and high reproduction rates. It is the case of brook trouts, an alien salmonid from North America, in the naturally fishless alpine lakes of the Gran Paradiso National Park [41]. This species has been introduced in the GPNP in the 1960s' due its fitness to extremely cold environments [32]. Because of the brook trout broad ecological valence, very soon the species established reproductive populations in the stocked lakes. Following the fishery ban in the 1970s', the brook trout found no more obstacles to its spread. Recent studies confirmed that the presence of brook trout in the alpine lakes of GPNP have the same strong ecological impact as observed in other mountain regions [23], with a dramatic impact on the macroinvertebratcs and zooplanktou communities, and on *Rana temporaria*. The heavy impact on biodiversity in high altitude alpine lakes is the motivation for an eradication campaign, undertaken by the GNPN within the EU financed LIFE+ BIOAQUAE (Biodiversity Improvement of Aquatic Alpine Ecosystems) project [56]. The aim here is the elimination of the threat represented by the invasive species, with strict requirements on the final state of the ecological system.

3   Wild-Boar Control Through Dynamical System Theory

Although in Italy wild boars cannot be considered as alien, in the Alta Murgia National Park they exhibit the typical behaviour of an invasive species, with a strong negative impact on agriculture. Due to the hunting ban in the site they spread fast not having practically competitors or predators with the sole exception of wolves. As already pointed out, wolves represent a protected species, and even if the Alta Murgia site extension and vegetation is not suited for hosting a viable and stable wolf population, their migration to Alta Murgia from nearby sites raises a conservation issue. The abatement program should take into account that the size of the wild boars population surviving the abatement has to be large enough to sustain the small population of wolves. Moreover, any conservation policy has to be established at a metapopulation level: functional connection areas (corridors) connecting Alta Murgia site (patch) to larger sites (patches) may allow the wolf-wild boar pair conservation [48]. Under the assumptions related to the specific case study, we derived a two-patch model for the wild boar and wolf population dynamics:

$$
\begin{cases}
\dfrac{dn}{d\tau} = \epsilon \left[r\, n \left(1 - \dfrac{n}{k} \right) \left(\dfrac{n}{A} - 1 \right) - a_1\, n\, p_1 - q\, E\, n \right], \\[2ex]
\dfrac{dp_1}{d\tau} = d_2\, p_2 - d_1\, p_1 + \epsilon \left[-(\mu + q\, E)\, p_1 + e\, a_1\, n\, p_1 \right], \\[2ex]
\dfrac{dp_2}{d\tau} = d_1\, p_1 - d_2\, p_2 + \epsilon \left[-\mu\, p_2 \right],
\end{cases}
\tag{1}
$$

where p_1 and p_2 represent the density of the wolf population (predator) in Alta Murgia Park (denoted as patch1) and in the Dauni Mountains site (denoted as patch2) respectively. The state variable n is the density of the wild-boar (prey) population in the Alta Murgia Park. The presence of wild-boar in patch2 is considered negligible for the aims of the present study since experimental field data indicate that the wild boar population are present in the Dauni Mountains at very low densities and no movements between patches was observed. The prey population is assumed to grow according an Allee law: r is the intrinsic growth rate, k is the carrying capacity and A is the Allee threshold, with $A < k$. The control parameter E is related to the abatement level of wild boars decided by the management authority of the Alta Murgia Park. Wild-boars are sedentary inside their living patch so that no migration is assumed from one patch to the other and vice versa. On the contrary, wolves move from patch1 (Alta Murgia) to patch2 (Dauni Mountains) at a constant rate of d_1 and from patch2 (Dauni Mountains) to patch1 (Alta Murgia) at a constant rate of d_2. The parameter a_1 is the predation coefficient, μ is the wolf death rate, and q is the catchability coefficient. Finally, ϵ is a small positive parameter, i.e. $\epsilon << 1$, τ is the fast time scale and $t = \tau \epsilon$ is the slow time scale: migration between patches is assumed to occur at a fast time scale in comparison with the local predator-prey dynamics. Since two different time scales are involved in the model, one can benefit of the use of the aggregation method [1] and derive a reduced model describing the dynamics of the total prey and predator densities n and p [35]:

$$\begin{cases} \dfrac{dn}{dt} = n \left[r \left(1 - \dfrac{n}{k} \right) \left(\dfrac{n}{A} - 1 \right) - a_1 d\, p - q\, E \right], \\[2mm] \dfrac{dp}{dt} = p \left[-\mu + e\, a_1 d\, n - q\, E\, d \right], \end{cases} \tag{2}$$

where $p = p_1 + p_2$ is the total amount of wolves and the positive parameter d is given by $d = \dfrac{d_2}{d_1 + d_2}$ and can be interpreted as the *relative migration rate* from patch2 to patch1, i.e. the ratio between the migration rate d_2 and the total migration rate $d_1 + d_2$. The parameter d can hence be considered as an interesting ecological indicator since it can be indirectly related to the existence of ecological corridors between patches. Theoretical results [1] ensure that if system (2) is structurally stable—as in this case—and the parameter ϵ is small enough, the dynamics of (2) is a good approximation of the dynamics of the global variables in (1). Because of their relevance in the ecological problem we are interested in, both the parameters d and E have been considered as bifurcation parameters and the following numerical values have been assigned to the other parameters: $a_1 = 1.108 \times 10^{-1}$; $r = 4.84 \times 10^{-2}$; $e = 2.8 \times 10^{-2}$; $\mu = 0.12$; $k = 120$; $A = 0.6182$; $q = 1$ [35].

By applying dynamical systems and bifurcation theory, qualitative results on the existence and stability properties of the attractors of model (2) as well as qualitative insights on the possible ecological scenarios obtained with varying some ecologically meaningful parameters are obtained in [35]. Theoretical results suggest

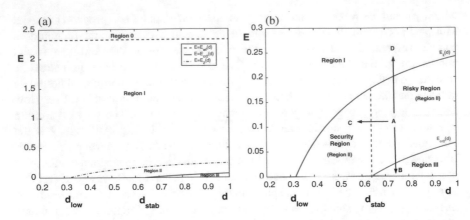

Fig. 1 The aggregated model (2). (**a**) Bifurcation diagram in the plane (d, E). The other parameter values are assigned as specified in the text. The curves $E = E_{wf}$, $E = E_2(d)$ and $E = E_{cr2}$ divide the plane in four regions: (i) Region 0 in which both the wolf and the wild-boar population will be extincted; (ii) Region I in which wolf population will be extincted in any case; (iii) Region II in which, according to the initial conditions, either wolf extinction or wolf survival via a stable coexistence equilibrium is expected (iv) Region III in which, according to the initial conditions, (**a**) wolf extinction or (**b**) wolf survival through small amplitude periodic oscillations in the strict neighboring of the Hopf bifurcation line $E = E_{cr2}(d)$ are expected. (**b**) A detail of the bifurcation diagram in the subfigure which specifically focuses on Region II in which coexistence between wolf and wild-boar population can be ensured through a stable coexistence equilibrium. Theoretical results suggests that from the point of view of the control strategies, Region II can be divided in two different subregions: a *Security Region* and a *Risky Region*

that survival of the wolf population in the Alta Murgia Park can be obtained only through wild boar and wolves coexistence. In this regard, a careful combination of processes as *control*—through proper planning programs—and *migration*—through the existence of suitable ecological corridors—must be used in order to properly control the wild-boar population while preserving wolf population from extinction.

The two-parameter bifurcation diagram in Fig. 1a, that depicts the different dynamical scenarios as functions of the two ecologically relevant parameters d and E, indicates in fact that (i) a threshold value d_{low} for the parameter d exists such that too low values of d, i.e. $d < d_{low}$, can lead either to the wolves extinction or to wild boars and wolves extinction independently of the value of the control parameter E; (ii) too high values of the control parameter E, i.e. $E > E_{wf}$ lead both the species toward extinction, independently of the value of the parameter d (iii) wolves survival is possible only in Region II in which coexistence between wolf and wild-boar populations can be ensured through a stable coexistence equilibrium or in Region III where—in the strict neighboring of the Hopf bifurcation line, coexistence can be obtained via a stable small amplitude limit cycle whereas away from the Hopf line, the amplitude of oscillations tends to increase. Since environmental fluctuations could lead to the extinction of the wolf populations, large amplitude oscillations might be prevented and Region II represents hence the most desirable situation in terms of a sustainable wild-boar and wolf management in the Alta Murgia Park. Moreover, from the point of view of the control strategies, Region II can be divided in two different subregions: a *Security Region* and a *Risky Region*, Fig. 1b. The *Security Region* is characterized by *sufficiently low E* values and *intermediate d-* values, i.e. $d_{low} < d < d_{stab} \ \land \ 0 < E < E_2$. It represents the 'perfect' region since it allows a sure monitoring of the control strategies: for a fixed value of d in the Security Region, coexistence can be ensured keeping the value of E below E_2. For E tending toward E_2, the wolf population level progressively decreases until it approach zero entering in Region I. In this case, a careful monitoring of the control can avoid wolf extinction by lowering E below the E_2 thresholds. These findings give hence some practical warnings for the specific case under study. In primis, either too low d values and high E values have to be avoided. Moreover *sufficiently low* values of the control parameter E have to be combined with *intermediate* values of the migration parameter d in order to ensure the coexistence between the wolf and wild-boar populations and hence the survival of the wolves in the Alta Murgia Park.

Therefore, the key warning from the model is that a very careful combination of control through proper planning programs and migration processes among patches of habitats through the existence of suitable ecological corridors must be used in order to properly limit the wild boar population while preserving wolves from extinction.

4 Wild-Boar Control Through Z-type Dynamic Approach

The Z-type dynamic method is an error-based dynamic approach in which the key element is the design formula which ensures that each element of the error function converges to zero exponentially. Since the Z-type control laws have exponential convergence performances, the convergence time can be estimated and populations will converge to the desired state in a fast or predefined rate. This feature enables the Z-type dynamic method to be particularly efficient for specific time limited applications. We describe the application of Z-type control to the aggregated model of the wolf-wild boars dynamics [36]:

$$\begin{aligned} \dot{n} &= n\big[H(n) - \beta p\big] \\ \dot{p} &= pF(n) \end{aligned} \tag{3}$$

with $H(n) = r\big(1 - \frac{n}{k}\big)\big(\frac{n}{a} - 1\big) - q\,E$; $F(n) = -\mu + e\,\beta\,n - q\,E\,d$ and $\beta = a_1\,d$. As in the previous section, $n = n(t)$ and $p = p(t)$ denote the total population densities of prey (wild-boars) and predator (wolves) respectively.

The Z-type control laws have the effect to change the dynamics of the *uncontrolled* model (3)—that may admit the predator-prey extinction equilibrium $P_0 = (0, 0)$, the predator's extinction equilibria $P_i = (n_i, 0)$, and the coexistence equilibrium $P = (n^*, p^*)$—as to make the prey and predator populations converge exponentially to a desired state $P^* = (n_d, p_d)$.

The Z-control is successful if the error between the actual output of the system and its desired state approaches zero, namely if

$$e(t) = n(t) - n_d(t) \to 0.$$

where $n_d = n_d(t)$ is the desired state for the prey population. This goal can be achieved by forcing the error function $e(t)$ to converge to zero exponentially, namely requiring that the error function verifies the differential equation

$$\dot{e}(t) = -\lambda\, e(t). \tag{4}$$

Equation (4) is also named *design formula* and the strictly positive parameter λ is the *design parameter* which indicates the convergence rate. The control procedure by the Z-type dynamic method is hence based on the following two steps: (i) define the error function(s); (ii) use the design formula whenever necessary to obtain an explicit expression for the input $u(t)$. These two steps will be concretely applied for the indirect control of the prey population of model (3) as follows.

By pursuing an indirect control on the prey population, model (3) becomes:

$$\begin{aligned}
\dot{n} &= n\big[H(n) - \beta p\big] \\
\dot{p} &= p\big[F(n) - u_{pred}\big]
\end{aligned} \tag{5}$$

where $u_{pred} = u_{pred}(t)$ denotes the indirect control variable for the prey population, acting on the predator dynamics, that can be both a positive or a negative quantity, [60]. The choice of controlling the prey population through a modified mortality of the predator is of interest in cases where the local protection status (e.g., a national park as in the case of Alta Murgia) prevents massive hunting on the prey (the wild boar), while targeted removal (and possibly transplant) and/or introduction of predators from other parks can be effectively managed by the park personnel. Following the key steps recalled above and applying the indirect Z-control procedure, we obtain the following system:

$$\begin{aligned}
\dot{n} &= n\big[H(n) - \beta p\big] \\
\dot{p} &= f(t; n, p)
\end{aligned} \tag{6}$$

with

$$f(t; n, p) = \frac{\sum_{k=0}^{5} C_k(p)\, n^k}{k^2\, a^2\, a_1\, d\, n}, \tag{7}$$

with the coefficients $C_k(p)$ given by:

$$C_5 = 3\,r^2;$$
$$C_4 = -5\,r^2\,(a+k);$$
$$C_3 = r\left[2\,r(a^2+k^2)+2\,a\,k(\lambda+4\,r)-a\,k(4\,a_1\,d\,p+3\,E\,q)\right];$$
$$C_2 = a\,k\,r\,(3\,a_1\,d\,p+2\,E\,q-2\lambda-3\,r)\,(a+k);$$
$$C_1 = k^2 a^2\left[(\lambda+r-a_1\,d\,p)(\lambda+r-a_1\,d\,p-E\,q)-E\,q\,\lambda\right];$$
$$C_0 = -k^2 a^2\left(\lambda^2\,n_d+2\,\lambda\,\dot{n}_d+\ddot{n}_d\right).$$

We refer to model (6) as to the *Z-controlled system* and, in the next, we specifically focus on the case $n_d(t) = n_d$, with n_d strictly positive constant.

Since $n_d = constant$, (7) reduces to

$$f(n,p) = \frac{n\left[\left(r\left(1-\frac{n}{k}\right)\left(\frac{n}{a}-1\right)-q\,E-a_1\,d\,p+\lambda\right)^2+\dot{H}(n)\right]-\lambda^2 n_d}{a_1\,d\,n},$$

with

$$\dot{H}(n) = \frac{n\,r}{k\,a}(k+a-2n)\left[r\left(1-\frac{n}{k}\right)\left(\frac{n}{a}-1\right)-q\,E-a_1\,d\,p\right].$$

Moreover,

$$u_{pred} = -\mu+e\,a_1\,d\,n-q\,Ed-\frac{f(n,p)}{p(t)}. \tag{8}$$

In the following, we use the Z control approach to modify two different dynamical scenarios of the uncontrolled model (3): (i) coexistence through a stable equilibrium P_e; (ii) coexistence through self-sustained oscillations surrounding the unstable equilibrium P_e.

As a first step, we fix the E and d parameter values so that, in the uncontrolled model (2), populations coexistence is reached by the mean of a stable equilibrium. We hence choose $E = 0.16$, $d = 0.6446$ and set the other model parameters as $a_1 = 0.1108$, $r = 0.0484$, $e = 0.0280$, $\mu = 0.12$, $k = 120$, $a = 0.6182$, $q = 1$. Basing on field-data, the chosen initial conditions are: $n(0) = 14$, $p(0) = 0.0563$, [35]. As a consequence, for the uncontrolled model (3), system dynamics tends towards a stable coexistence equilibrium $P_e = (111.545, 6.325)$, Fig. 2.

For the chosen parameter values and in respect to the initial conditions $P_0 = (n_0, p_0) = (14, 0.0563)$, the parameter λ has to be chosen such that $0.1240 \leq \lambda \leq 0.6224$ in order the Z-control approach to be ecologically meaningful for the aggregated model of the wolf-wild boar dynamics.

Figure 3 shows the basin of attraction of the Z-controlled equilibrium P^* for increasing values of the parameter $\lambda := 0.2; 0.5$. In both these cases, the initial point $P_0 = (14, 0.0563)$ is inside the ecological basin of attraction of P^*, that considerably enlarges for increasing values of λ.

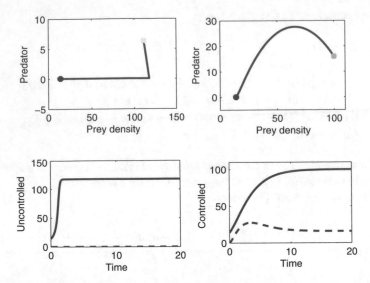

Fig. 2 The aggregated model of the wolf-wild boar dynamics [35]: Scenario I. First row: Phase-plane curves of predator versus prey for the aggregated model without control (left) and with control (right). The design parameter of the Z-controlled model is $\lambda = 0.5$. For both the models the initial conditions are $n(0) = 14$, $p(0) = 0.0563$. The parameters E and d are chosen as $E = 0.16$, $d = 0.6446$. The other system parameters are chosen as $a_1 = 0.1108$, $r = 0.0484$, $e = 0.0280$, $\mu = 0.12$, $k = 120$, $a = 0.6182$, $q = 1$. The uncontrolled system dynamics asymptotically reach the stable equilibrium $P_e = (111.545, 6.325)$ while the Z-control forces the system dynamics towards the desired equilibrium $P^* = \left(100, 15.9124\right)$. In the phase-plane, the initial point $P_0 = (n(0), p(0))$ is plotted as a red full circle; the uncontrolled equilibrium P_e is plotted as a yellow full circle; the Z-controlled equilibrium P^* is plotted as a green full circle. Second row: Time-dependent behavior of the prey and predator populations of the aggregated model (Color figure online)

Now we fix the E and d parameter values so that, in the uncontrolled model (2), populations coexistence is reached by the mean of self-sustained oscillations. We hence choose $E = 0.0537$, $d = 0.9$ and set the other model parameters as $a_1 = 0.1108$, $r = 0.0484$, $e = 0.0280$, $\mu = 0.12$, $k = 120$, $a = 0.6182$, $q = 1$. As in the previous example, the chosen initial conditions are: $n(0) = 14$, $p(0) = 0.0563$. As a consequence, for the uncontrolled model (2), system dynamics tends towards a stable limit cycle surrounding the unstable equilibrium $P_e = (60.268, 22.766)$, Fig. 4.

Hence, for the parameter values chosen in this section and in respect to the initial conditions $P_0 = (n_0, p_0) = (14, 0.0563)$, the parameter λ has to be chosen such that $0.5774 \leq \lambda \leq 1.8374$ for the effectiveness of the Z-control approach.

Figure 5 depicts the basin of attraction of the Z-controlled equilibrium P^* for increasing values of the parameter $\lambda := 0.6; 1; 1.8$. In all the three cases the initial point $P_0 = (14, 0.0563)$ is inside the ecological basin of attraction of P^* and, increasing values of λ, the ecological basin of attraction of the Z-controlled

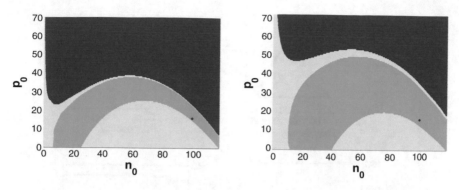

Fig. 3 The Z-controlled aggregated model of the wolf-wild boar dynamics: Scenario I. Basin of attraction of the Z-controlled equilibrium P^*, for different values of the parameter λ. The green region (dark grey in the printed version) represents the ecological basin of attraction of the Z-controlled equilibrium P^*. The yellow region (light grey in the printed version) represent the theoretical but not ecologically feasible basin of attraction of the Z-controlled equilibrium P^*. The blue region (black in the printed version) represents the set of initial conditions for which the Z-control approach fails. The different panels are distinguished by the different numerical values assigned to parameter λ: left, $\lambda = 0.2$; right, $\lambda = 0.5$. In each panel, the stable equilibrium P^* is marked through a red filled circle (Color figure online)

equilibrium P^* considerably enlarges. In conclusion, a population model with a Z-control is shown to have two positive effects on the dynamics, that turn out to be interesting in the perspective of ecosystem services management. The first one is the stability: for some population models, the introduction of a control can lead to the same equilibria of the original uncontrolled model, but with additional stability properties. A further advantage is the capability to generate new, stable equilibria, by construction. That might be interesting by an ecological point of view in the case the new equilibria turn out to be more robust with respect to external, uncontrolled, disturbance factors.

5   Wild-Boar Control Through Optimal Control Theory

To illustrate how the optimal control theory can be used to face the wild-boars problem, we start by considering the general case of an invasive species population settled on a small region and assume that the main focus of the management activity is to reduce such a population to a certain fixed level, at the lowest possible cost. Since in small regions the species spatial distribution has little impact on the management outcomes, we focus only on the temporal features of the management problem.

With the assumption that it is uniformly distributed and well-mixed in space, the dynamics of the invasive species population can be described by the following

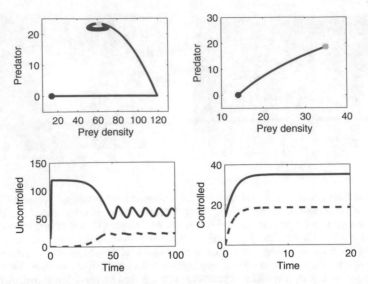

Fig. 4 The aggregated model of the wolf-wild boar dynamics [35]: Scenario II. First row: Phase-plane curves of predator versus prey for the aggregated model without control (left) and with control (right). The design parameter of the Z-controlled model is $\lambda = 1$ and the initial conditions are $n(0) = 14$, $p(0) = 0.0563$. The parameters E and d are chosen as $E = 0.0537$, $d = 0.9$. The other system parameters are chosen as $a_1 = 0.1108$, $r = 0.0484$, $e = 0.0280$, $\mu = 0.12$, $k = 120$, $a = 0.6182$, $q = 1$. The uncontrolled system dynamics asymptotically reach a stable limit cycle surrounding the unstable equilibrium $P_e = (60.268, 22.766)$ while the Z-control forces system dynamics towards the desired equilibrium $P^* = (35, 18.577)$. In the phase-plane, the initial point $P_0 = (n(0), p(0))$ is plotted as a red full circle; the uncontrolled equilibrium P_e is plotted as a yellow full circle; the Z-controlled equilibrium P^* is plotted as a green full circle. Second row: Time-dependent behavior of the prey and predator population for the aggregated model (Color figure online)

ordinary differential equation:

$$\dot{u} = u f(u) - u \left(\mu(t) E \right)^q , \tag{9}$$

where $u = u(t)$ is the density of the invasive population we aim to control and $\dot{u} = \frac{du}{dt}$. Equation (9) is a balance between the density-dependent population growth term $u f(u)$ and the effect on the population of the control actions, due to the term $-u \left(\mu(t) E \right)^q$, where $E = E(t)$ is the control function. Unlike some other models of the dynamics of harvested populations [34], in (9) the control effort impacts the dynamics not linearly, with the functional form $\left(\mu(t) E \right)^q$ that was introduced in [2–5, 44] to model the more realistic circumstance that the control efforts exhibit decreasing marginal returns on investments. This means that if one applies additional control effort when the control effort is already large, this produces a smaller incremental benefit compared to when the control effort is low.

The proportional reduction of the invasive population by the control effort, is modulated by the (strictly) positive scaling function $\mu(t) > 0$, which accounts

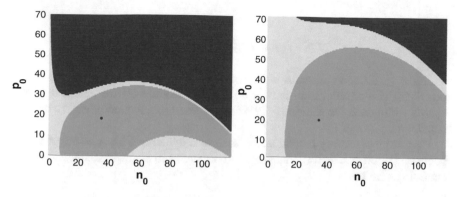

Fig. 5 The Z-controlled aggregated model of the wolf-wild boar dynamics: Scenario II. Basin of attraction of the Z-controlled equilibrium P^*, for different values of the parameter λ. The green region (dark grey in the printed version) represents the ecological basin of attraction of the Z-controlled equilibrium P^*. The yellow region (light grey in the printed version) represents the portion not ecologically meaningful of the theoretical basin of attraction of P^*. The blue region (black in the printed version) represents the set of initial conditions for which the Z-control approach fails. The different panels are distinguished by the different numerical values assigned to parameter λ: (left) $\lambda = 0.6$; (right) $\lambda = 1.8$. In each panel, the stable equilibrium P^* is marked through a red filled circle (Color figure online)

for the control effectiveness, and by the parameter $q \in \mathbb{Q} \cap \left[\frac{1}{2}, 1\right)$, that is the diminishing returns parameter. More precisely, low values of the parameter q indicate control actions that are not cost-effective at high intensity since the related marginal returns decrease very quickly. Bounds on q guarantee both the convexity of the Hamiltonian ($q < 1$) and the uniqueness of the optimal solution ($q \geq 1/2$).

To attack the problem, we propose to investigate an optimality system—the state-control model which is alternative to the classical state-costate optimality system. The underlying idea—developed in details in [5, 44]—is to couple the optimal control theory and the phase-plane analysis with the aim to derive analytical insights on the optimal strategy for the control of the invasive species.

We suppose that the invasive species population is initially at the density $u(0) = u_0$ and assign $u(T) = u_T$, with $0 < u_T < u_0$, meaning that the invasive species has to be reduced to the threshold density u_T at exactly $t = T$, where T represents the length of the project.

We formulate an optimal control problem with fixed terminal time as follows: given $g(t, u, E)$ a continuously differentiable function (in all three variables), which represents the project cost at the time t, we seek $E(t)$ such that the following expression is minimized

$$\min_{E \in U_{a,b}} \int_0^T g(t, u(t), E(t)) \, dt, \tag{10}$$

where

$$U_{a,b} = \{E \in L^1(0, T) : a \leq E \leq b\},$$

with $a, b \geq 0$ fixed constants, is the set of positive bounded Lebesgue integrable functions. The minimization is over all the admissible controls that is the set of control which steer the solution of (9) from the fixed initial condition u_0 to the fixed condition u_T at $t = T$.

First order necessary conditions, in case of a dynamic with Alee effect as wild boar population, can be formulated in terms of state and control variables. The state-control system reads:

$$\dot{u} = r u \left(\frac{u}{k_0} - 1\right) \left(1 - \frac{u}{k}\right) - u \mu^q E^q, \tag{11}$$

$$\dot{E} = A_1 (2u - k - k_0) u E, \tag{12}$$

for $0 \leq t \leq T$, where $A_1 = \dfrac{r}{k k_0 (1 - q)}$ is a positive constant because of the assumption $0 < q < 1$. We suppose that the assigned conditions for the state variable u are $u(0) = u_0$ and $u(T) = u_T$, with $k_0 < u_T < u_0 \leq k$, and we suppose $u_0 < k$ and we will use the same numerical values considered in [35, 36]: $r = 0.0484, k = 120, k_0 = 0.6182$ which are relative to the wild boar population in Alta Murgia Natural Park. We choose the bait effectiveness parameter such that $\mu = 1$ and we set the diminishing returns parameter so that $q = 0.64$. We can hence show that the optimal project length assumes a finite value T^* corresponding to the curve of minimal efforts, as stated in the following theorem [5]

Theorem 1 *Let $\mu > 0$ be a fixed constant, $\bar{T} > 0$ arbitrarily large, and consider the control set $U_{0,b} = \{E \in L^1(0, T) : 0 \leq E \leq b, 0 \leq T \leq \bar{T}\}$, with $b \geq \dfrac{1}{\mu} \left(\dfrac{r (k - k_0)^2}{2 (1 - q) k_0 k}\right)^{1/q}$, and $q \in \mathbb{Q}$, with $\frac{1}{2} \leq q < 1$. If $k_0 < u_T < u_0 < k$, the minimization problem*

$$\min_{(E,T) \in U_{0,b} \times [0,\bar{T}]} \int_0^T \mu E(t) \, dt \tag{13}$$

subject to

$$\dot{u} = r u \left(\frac{u}{k_0} - 1\right) \left(1 - \frac{u}{k}\right) - u \mu^q E^q \quad 0 \leq t \leq T, \ u(0) = u_0, \ u(T) = u_T, \tag{14}$$

has the optimal solution (E^*, T^*), where

$$T^* = \frac{1}{A_1} \log \left[\left(\frac{u_T}{u_0} \right)^B \left(\frac{u_0 - k_0}{u_T - k_0} \right)^C \left(\frac{k - u_T}{k - u_0} \right)^D \right], \tag{15}$$

and

$$E^*(t) = \frac{1}{\mu} \left[\frac{r}{1 - q} \left(\frac{u^*}{k_0} - 1 \right) \left(1 - \frac{u^*}{k} \right) \right]^{1/q}, \quad t \in [0, T^*]. \tag{16}$$

Moreover, the optimal solution $u^*(t)$ satisfies

$$\frac{(u - k_0)^C}{u^B (k - u)^D} = \frac{(u_{T^*} - k_0)^C}{u_T^B (k - u_{T^*})^D} e^{A(T^* - t)}, \quad t \in [0, T^*]. \tag{17}$$

In conclusion, investigating the wild-boar problem using the optimality state-control system has allowed us to give the analytic expressions for the minimum control effort. At this regard, we want to stress that understanding the optimal time horizon of an abatement or eradication project is of utmost importance in terms of strategic planning, since it improves the long term sustainability of nature conservation actions.

6 A Future Perspective: Optimal Spatio-Temporal Resources Allocation

The introduction of space into the optimal control model is aimed to account for the spatial features of alien species invasion and of environments that change under the pressure from external drivers. The treatment of optimal control problems governed by partial differential equations is less straightforward: generally speaking, for most nonlinear PDEs, a solution with continuous derivatives does not exist and the approximation of the solution for the state variable can be given in weak sense in a Sobolev space [25, 39, 40, 43]. Examples of PDEs optimal control problems of ecological interest are illustrated in [27, 37].

Here, we describe the optimal control problem of a PDE reaction diffusion model with a logistic growth, introduced in [2, 6]. The model describes a complex and realistic situation by including a control term that has Holling-II type behavior, a budget constraint and the habitat suitability function which represents the suitability of habitat for a given species based on known affinities with environmental parameters. High-quality habitat may provide high carrying capacity and support higher rates of growth, survival, or reproduction for a given species, whereas low-quality or unsuitable habitat may have little or no carrying capacity [9]. The goal is to minimize the environmental damage over time at the minimum cost, in terms of the resources allocated to the species harvesting. Assuming that the environmental damage has a cost which increases with the presence of the invasive species, the objective function assumes the following form:

$$J(E) = \int_{\Omega} e^{-\delta T} v(\mathbf{x}) n(\mathbf{x}, T) \, d\mathbf{x}$$

$$+ \int_{\Omega \times [0,T]} e^{-\delta t} \left(c \left(\frac{E(\mathbf{x}, t)}{B} \right)^3 + E^2(\mathbf{x}, t) + \omega(\mathbf{x}, t) n \right) d\mathbf{x} \, dt,$$

where $n(\mathbf{x}, t)$ represents the population density at time $t \in [0, T]$ and (vector) position $\mathbf{x}$ in an open bounded domain $\Omega \subset \mathbb{R}^2$, with a smooth boundary $\partial \Omega$. The parameter ω corresponds the cost due to the environmental damage, v is a weight for the final population density, $\delta \in (0, 1)$ is the discount factor. In the following, we assume $v \in L^{\infty}(\Omega)$, $v(\mathbf{x}) \geq 0$ for any $\mathbf{x} \in \Omega$, and $\omega \in L^{\infty}(\Omega \times [0, T])$, $\omega(\mathbf{x}, t) > 0$ over $\Omega \times [0, T]$. The term

$$\int_{\Omega \times [0,T]} e^{-\delta t} \left(\frac{E(\mathbf{x}, t)}{B} \right)^3 d\mathbf{x} \, dt,$$

where $B > 0$, accounts for the budget constraint, introduced as a penalty term with weight $c > 0$ in the objective function or it can be neglected by setting $c = 0$. We search for a control

$$E^* \in \mathcal{U} = \{E \in L^{\infty}(\Omega \times [0, T]) : 0 \leq E(\mathbf{x}, t) \leq B \text{ for all } (\mathbf{x}, t) \in \Omega \times [0, T]\}$$

such that $J(E^*) = \min_{E \in \mathcal{U}} J(E)$, subjects to the state equation

$$\frac{\partial n}{\partial t}(\mathbf{x}, t) - D \Delta n(\mathbf{x}, t) = r \, n(\mathbf{x}, t) \left(\rho(\mathbf{x}) - \frac{n(\mathbf{x}, t)}{k} \right) - \frac{\mu \, n(\mathbf{x}, t) \, E(\mathbf{x}, t)}{1 + h \mu n(\mathbf{x}, t)}, \quad (18)$$

$$n(\mathbf{x}, 0) = n_0(\mathbf{x}), \text{ on } \Omega, \quad \nabla n(\mathbf{x}, t) \cdot \hat{\mathbf{n}} = 0, \text{ on } \partial \Omega \times [0, T], \quad (19)$$

where $D > 0$ represents the constant diffusion coefficient and $\hat{\mathbf{n}}$ is the outward normal vector on $\partial \Omega$. The term $\rho(\mathbf{x})$ represents the habitat suitability function which is bounded $\rho(\mathbf{x}) \in [0, 1]$ for each $\mathbf{x} \in \Omega$; that implies $\rho \in L^{\infty}(\Omega)$ and $\|\rho\|_{L^{\infty}(\Omega)} \leq 1$. Under the assumption that $n_0 \in H^1(\Omega)$ the usual Sobolev space (with its dual $H^1(\Omega)^*$), the solution of the problem (18)–(19), is to be considered in week sense that is $n \in L^2(0, T; H^1(\Omega))$ such that $\frac{\partial n}{\partial t} \in L^2(0, T; H^1(\Omega)^*)$ and

$$n(\mathbf{x}, 0) = n_0(\mathbf{x}), \quad \text{on } \Omega, \quad (20)$$

$$\int_{\Omega} \frac{\partial n}{\partial t} \chi \, d\mathbf{x} + \int_{\Omega} D \nabla n \cdot \nabla \chi \, d\mathbf{x} = \int_{\Omega} (r\rho - f(n, E)) n \chi \, d\mathbf{x}, \quad (21)$$

for almost every $t \in [0, T]$ and for any test function $\chi \in H^1(\Omega)$ [6].

The optimality conditions yield the following boundary value system

$$\frac{\partial n}{\partial t} = D\,\Delta n + r\,n\left(\rho - \frac{n}{k}\right) - \frac{\mu\,n\,E}{1 + h\,\mu\,n},$$

$$\frac{\partial \lambda}{\partial t} = -D\,\Delta\lambda + (\delta - r\,\rho)\,\lambda + \frac{2\,r\,n\,\lambda}{k} + \frac{\mu\,\lambda\,E}{(1 + h\,\mu\,n)^2} - \omega,$$

$$n\,(\mathbf{x}, 0) = n_0\,(\mathbf{x}),\ \text{on}\ \Omega, \quad \nabla n \cdot \hat{\mathbf{n}} = 0,\ \text{on}\ \partial\Omega \times [0, T],$$

$$\lambda\,(\mathbf{x}, T) = \nu\,(\mathbf{x}),\ \text{on}\ \Omega, \quad \nabla\lambda \cdot \hat{\mathbf{n}} = 0,\ \text{on}\ \partial\Omega \times [0, T],$$

equipped with

$$E(\mathbf{x}, t) = \min\{\varphi(n(\mathbf{x}, t), \lambda(\mathbf{x}, t)), B\},\ \text{on}\ \Omega \times [0, T], \tag{22}$$

where

$$\varphi(n, \lambda) = \begin{cases} \dfrac{B^3}{3c}\left(\sqrt{1 + \dfrac{3\,c\,\mu\,n\,\lambda}{B^3\,(1 + h\,\mu\,n)}} - 1\right), & \text{if}\ c > 0, \\[4ex] \dfrac{\mu\,n\,\lambda}{2\,(1 + h\,\mu\,n)}, & \text{if}\ c = 0. \end{cases} \tag{23}$$

The approximation process of the whole dynamics assumes a priori semi-discretization in the space variable performed by Finite Element method, applied with linear finite elements; the resulting procedure leads to an ordinary differential system in the time variable, which can be numerically integrated splitting the flow into two parts: the diffusive and the reaction term and then by composing the approximated flows (see [6, 7, 17, 20, 21]).

In [6] the approach is applied on two hypothetical examples that are representative of two realistic situations. In the first one, a triangular spatial domain is considered: it represents a natural zone which is destined to be employed as a parking area. The real domain is shown in Fig. 6. On 2013, the spatial domain was occupied by low vegetation and some trees delimited the area along its wall. At a later time, on 2015, the vegetation was cut down and the area owners covered the soil by gravel; in this way, they made the spatial domain less suitable for vegetation growing. On 2017, it front of vegetation advanced spreading along edges and corners, which correspond to the parts that are not easily reachable by cars in the parking. We assume that the owners have a budget constraint to be employed in order to keep the area clean; then, the interest is focused on the best budget allocation both in space and in time, for parking area maintenance. We show the results of some simulations on the spatial area simplified as a triangular domain Ω, with measure $|\Omega| = 1/2$, where a mesh of 512 triangles and 289 nodes is built by

Fig. 6 The real domain is shown. The pictures on the left and in the middle date back to years 2013 and 2015, respectively; they are taken from *Map data: Google*. The picture on the right is dated 2017; it is taken from *Map data: Google, SPOT Image*

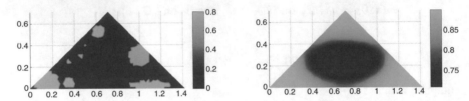

Fig. 7 On the left side: values of initial distribution defined as $n_0(x, y) = \alpha_i$ if $\left(\frac{x - X_i}{a_i}\right)^2 + \left(\frac{y - Y_i}{b_i}\right)^2 < 1$, for $i = 1 \ldots 8$. Parameters are chosen as $\alpha_i = 0.8$ for $i = 1, 2, 3, 5$, and $\alpha_i = 0.7$ for $i = 4, 6, 7, 8$, $X = [0.15, 0.35, 1.1, 1, 0.62, 0.5, 0.45, 0.35]$, $Y = [0.05, 0.05, 0.03, 0.3, 0.52, 0.4, 0.3, 0.25]$, $a = [0.11, 0.05, 0.2, 0.1, 0.06, 0.02, 0.02, 0.02]$, $b = [0.11, 0.05, 0.04, 0.1, 0.06, 0.02, 0.02, 0.02]$. On the right side: density distribution evaluated at $T = 20$ without control, i.e. setting $E = 0$ everywhere

means of `mesh2D` Matlab function. The parking area is represented by an ellipses inside the whole Ω. The suitability function is set at lower value $\rho(x, y) = 0.7$ at the zone occupied by cars (inside the ellipses); while, a higher value $\rho(x, y) = 0.9$ is set where the soil is sealed by gravel. The picture on the left side of Fig. 7 shows the initial distribution $n_0(x, y)$ which simulates the real situation in 2013. We suppose that the control action is planned over a time horizon up to $T = 1$ and we advance by temporal stepsize $\Delta t = 1/36$. Vegetation parameters are set at the values $r = 1.92$, $k = 1, \omega = 1, \mu = 9, \nu = 5 e^{-0.1}, h = 1, \delta = 0.1, D = 0.0052$.

We compare two different situations. First, we consider the case without penalty term ($c = 0$) for the budget $B = 1$; then we simulate the dynamics by assuming a penalty term with weight $c = 1$. In Fig. 8 the results of both simulations are shown: in absence of penalty term for budget constraint (on the first row of Fig. 8) it is evident that, at the end of the dynamics, vegetation reaches its minimum value and the eradication is achieved by an optimal effort $E_0^*(x, y, 1)$ which gets a value equal to $B = 1$ at several points of the spatial domain. When we introduce the penalty $c = 1$ on budget constraint (on the second row of Fig. 8), a small quantity of vegetation still persists in the lower angle on the left of the parking area Ω.

In the second example, we simulate an hypothetical invasion of alien fish population inside a lake. In this case, the algorithm we provide faces also a very complex geometry representing a realistic domain, where it may be reasonable to suppose very large spatial gradients for $\rho(x, y)$. To perform the simulations

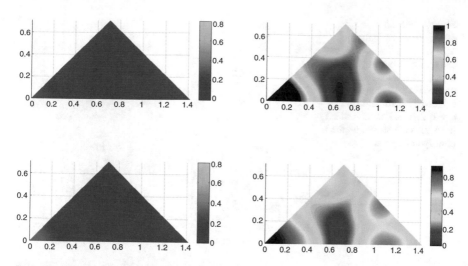

Fig. 8 First example: density distribution (on the left) and control effort (on the right) evaluated at $T = 1$. On the first row, the results are shown for $c = 0$ and $B = 1$. On the second row, the results corresponding to $c = 1$ and $B = 1$ are plotted

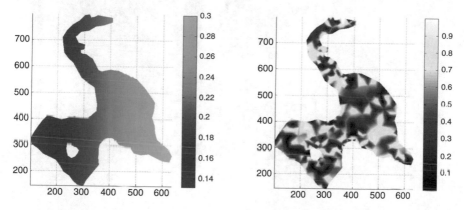

Fig. 9 Second example: scale of values for initial density distribution, which is defined as $n_0(x, y) = 0.2 x/633.9053 + 0.1$ (on the left side); scale for suitability values generated by means of Matlab built-in function rand (on the right side)

on the growth and diffusion of the fish population, we consider the spatial mesh consisting of 692 triangles and 427 nodes used in [22]. The discretization in time corresponds to a time step $\Delta t = 1/36$. We randomly generate the spatial distribution of habitat suitability $\rho(x, y)$ by using the built-in Matlab function rand (see Fig. 9, on the right). The parameters for fish population are set to $r = k = 1$, $\mu = 20$, $h = 1$, with a diffusivity coefficient $D = 0.05$; the parameters related to control dynamics are set as $\nu = 5 e^{-0.4}$, $\delta = 0.1$, $\omega = 1$. A given initial density of population is introduced into the lake (Fig. 9, on the left). We suppose to plan a control action over a time horizon with length $T = 4$ and we set a penalty term

Fig. 10 Second example: density of fish population $n(x, y, t)$ (on the left column) and optimal control $E(x, y, t)$ (on the right column). The dynamics is evaluated at different times: $t = 1, 2, 3, 4$. For each time t, the results are shown on the corresponding row

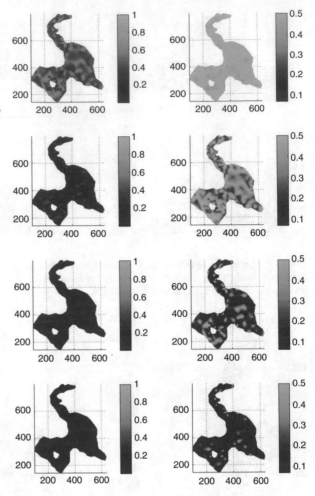

weighted by coefficient $c = 0.22$ for the budget constraint $0 \leq E \leq B = 0.5$. We start from a uniform condition for the population distribution, which corresponds to the maximum density $n_0(x, t) = 1$ and we evaluate the evolution dynamics at $t = 1, 2, 3, 4$: the results are shown in Fig. 10. We find that fish population tends to a final distribution $n(x, y, 4)$, with mean value 0.0012, integrated over the whole spatial domain. Moreover, in the same Fig. 10 it is evident that the maximum allowed value $E = B = 0.5$ is achieved in the areas where the sensitivity function gets its largest values. The optimal control tends to nullify the fish population almost everywhere except for some small areas where it gets its maximum allowed value.

7 Conclusions

The environmental and economical damage caused by invasive species has a significant cost, and any strategy designed to reduce their impact has to take into account all the constraints, either physical or budgetary. There is not a unique solution and, consequently, not a unique approach to solve the problem. The aim might be the reduction, as in the case of the wild boars, where a residual population is functional to the preservation of an endangered species, the wolf, or it might be the total elimination, as it is the case of the brook trouts in the alpine lakes and of the tree of heaven in a protected area, where the goal is the biodiversity conservation. Moreover, the context information and conditions, i.e. the patchiness of the habitat, the time horizon of the harvesting program, and finally the budget constraints, suggests a variety of approaches like the ones here considered.

The three approaches reported in the present paper represent a comprehensive, although not exhausting, compendium of model driven solutions to the invasive species control, which can be extended to several different species and adapted to many environmental conditions. The first approach, in the framework of the population dynamics, uses the dynamical system theory to analyze and suggests control strategies of an invasive species which feeds an endangered one. The approach identifies a region in the space parameters where the harvesting of the invasive species can be done while preserving at the same time the existence of the other species. The model is represented by a system of ODEs where some spatial information is embedded in the model parameters (habitat patchiness and migration). The second example is an error-based dynamic approach, again applied in time to a pair of species. The advantage of the approach is the possibility of setting the goals of the harvesting action and to control the dynamics and its equilibria through a specific design parameter influencing the stability properties of the solution. The third approach here shown, focus on just one species with the aim of minimize the effort required to eliminate it. As a result we obtain the curve of minimal effort to be allocated in the harvesting program, as well as its optimal duration. All the above solutions do not explicitly take the space into account. As a further development of the last approach we consider a nonlinear PDE model and the optimal control of an invasive species when we assume that there is a limit on the budget that can be allocated to the harvesting program. In this last case the model is described by a PDE reaction diffusion system with a logistic growth and the control term has a Holling-II type behavior. The approximation of the solution for the state variable is given in the weak sense in a Sobolev space. The approach has been applied on two case studies: the eradication of an invasive plant species in a parking area and the harvesting of an alien fish species in a mountain lake. On both cases we show that the optimal distribution of the effort in space and time ensure the local extinction of the invasive species, whereas an equal overall amount of budget, if uniformly allocated, lead to the persistence of the alien species. The models, the analytical studies on its properties, the numerical methods specifically designed as well as the applications, represent the steps of a long journey to come to a model driven solution of the ecological problem of the invasive species containment. A model driven approach to the problem would allow not only to solve the specific problem under investigation but also to perform a scenario analysis in environmental conditions which may vary due to external drivers (climate change, land use change, etc.). As a further development, we are currently working at the calibration of the model parameters for a concrete control problem in a protected area, where the biodiversity conservation is threatened by an invasive plant species. In-field data and remote sensing imagery will be integrated and assimilated into the model to provide both qualitative and quantitative suggestions for an optimal control strategy.

Acknowledgements This work has been carried out within the H2020 project 'ECOPOTEN-TIAL: Improving Future Ecosystem Benefits Through Earth Observations', coordinated by CNR-IGG (http://www.ecopotential.project.eu). The project has received funding from the European Union's Horizon 2020 research and innovation programme (grant agreement No 641762).

This work has been carried out within the LIFE Alta Murgia project (LIFE12 BIO/IT/000213—http://lifealtamurgia.eu/en/) coordinated by CNR-ISPA, titled Control and eradication of the invasive and exotic plant species *Ailanthus altissima* in the Alta Murgia National Park, funded by the European Commission under the LIFE Programme. D.L. research work has been performed under the auspices of the Italian National Group for Mathematical Physics (GNFM-INdAM).

References

1. Auger, P., Bravo de la Parra, R.: Methods of aggregation of variables in population dynamics. CR Acad. Sci. Sci. dela Vie **323**, 665–674 (2000)
2. Baker, C.M.: Target the source: optimal spatiotemporal resource allocation for invasive species control. Conserv. Lett. **10**, 41–48 (2017)
3. Baker, C.M., Bode, M.: Spatial control of invasive species in conservation landscapes. Comput. Manag. Sci. **10**, 331–351 (2013)
4. Baker, C.M., Bode, M.: Placing invasive species management in a spatiotemporal context. Ecol. Appl. **26**, 712–725 (2016)
5. Baker, C.M., Diele, F., Lacitignola, D., Marangi, C., Martiradonna, A.: Optimal control of invasive species through a dynamical systems approach. Nonlinear Anal. Real World Appl. **49**, 45–70 (2019)
6. Baker, C.M., Diele, F., Marangi, C., Martiradonna, A., Ragni, S.: Optimal spatiotemporal effort allocation for invasive species removal incorporating a removal handling time and budget. Nat. Resour. Model. **31**(4), e12190 (2018)
7. Blanes, S., Diele, F., Marangi, C., Ragni, S.: Splitting and composition methods for explicit time dependence in separable dynamical systems. J. Comput. Appl. Math. **235**(3), 646–659 (2010)
8. Bonnans, J.F., Varin, J.L.: Computation of order conditions for symplectic partitioned Runge-Kutta schemes with application to optimal control. Numer. Math. **103**, 1–10 (2006)
9. Brown, D.R., Stouffer, P.C., Strong, C.M.: Movement and territoriality of wintering Hermit Thrushes in southeastern Louisiana. Wilson Bull. **112**, 347–353 (2000)
10. Burnett, K., Pongkijvorasin, S., Roumasset, J.: Species invasion as catastrophe: the case of the brown tree snake. Environ. Resour. Econ. **51**, 241–254 (2012)
11. Casella, F., Vurro, M.: Ailanthus altissima (tree of heaven): spread and harmfulness in a case study urban area. Arboricult. J.: Int. J. Urban For. **35**(3), 172–181 (2013). https://doi.org/10.1080/03071375.2013.852352

12. Celesti-Grapow, L., Pretto, F., Carli, E., Blasi, C.: Flora vascolare alloctona e invasiva delle regioni d'Italia [Allochthonous and Invasive Vascular Flora of the Regions of Italy]. Casa Editrice Universitaria La Sapienza, Rome (2010)
13. Chalak, M., Pannell, D.J.: Optimising control of an agricultural weed in sheep-production pastures. Agric. Syst. **109**, 1–8 (2012)
14. Chalak-Haghighi, M., van Ierland, E.C., Bourdot, G.W., Leathwick, D.: Management strategies for an invasive weed: a dynamic programming approach for Californian thistle in New Zealand. N. Z. J. Agric. Res. **51**, 409–424 (2008)
15. Chyba, M., Hairer, E., Vilmart, G.: The role of symplectic integrators in optimal control. Optim. Control Appl. Methods **30**, 367–382 (2009)
16. Colautti Robert, I., MacIsaac, H.J.: A neutral terminology to define invasive species. Divers. Distrib. **10**(2), 135–141 (2004)
17. Diele, F., Marangi, C.: Positive symplectic integrators for predator-prey dynamics. Discrete Contin. Dynam. Syst. B **23**(7), 2661 (2017)
18. Diele, F., Marangi, C., Ragni, S.: Steady-state invariance in high-order Runge-Kutta discretization of optimal growth models. J. Econ. Dyn. Control **34**, 1248–1259 (2010)
19. Diele, F., Marangi, C., Ragni, S.: Exponential Lawson integration for nearly Hamiltonian systems arising in optimal control. Math. Comput. Simul. **81**(5), 1057–1067 (2011)
20. Diele, F., Marangi, C., Ragni, S.: Implicit-symplectic partitioned (IMSP) Runge-Kutta schemes for predator-prey dynamics. In: AIP Conference Proceedings, vol. 1479, no. 1. American Institute of Physics (2012)
21. Diele, F., Marangi, C., Ragni, S.: IMSP schemes for spatially explicit models of cyclic populations and metapopulation dynamics. Math. Comput. Simul. **100**, 41–53 (2014)
22. Diele, F., Garvie, M.R., Trenchea, C.: Numerical analysis of a first-order in time implicit symplectic scheme for predator prey systems. Comput. Math. Appl. **74**(5), 948–961 (2017)
23. Eby, L.A., Roach, W.J., Crowder, L.B., Stanford, J.A.: Effects of stocking-up freshwater food webs. Trends Ecol. Evol. **21**, 576–584 (2006)
24. European Commission Staff, Working Document Impact Assessment Accompanying the Document Proposal for a Council and European Parliament Regulation on the prevention and management of the introduction and spread of invasive alien species. http://eur-lex.europa.eu/legal-content/EN/TXT/?uri=CELEX:52013SC0321 (2013)
25. Evans, L.C.: Partial Differential Equations. American Mathematical Society, Providence (1998)
26. Ficetola, G., Bonardi, A., Mairota, P., Leronni, V., Padoa-Schioppa, E.: Predicting wild boar damages to croplands in a mosaic of agricultural and natural areas. Curr. Zool. **60**, 170–179 (2014)
27. Garvie, M.R., Trenchea, C.: Optimal control of a nutrient-phytoplankton-zooplankton-fish system. SIAM J. Control Optim. **46**(3), 775–791 (2007)
28. Guckenheimer, J., Holmes, P.: Nonlinear Oscillations, Dynamical Systems and Bifurcations of Vector Fields. Springer, New York (1997)
29. Guo, D., Zhang, Y.: Neural dynamics and Newton-Raphson iteration for nonlinear optimization. J. Comput. Nonlinear Dyn. **9**(2), 021016 (2014)
30. Hager, W.W.: Runge-Kutta methods in optimal control and the transformed adjoint system. Numer. Math. **87**, 247–282 (2000)
31. Hairer, E., Lubich, C., Wanner, G.: Geometric Numerical Integration. Structure-Preserving Algorithms for Ordinary Differential Equations, 2nd edn. Springer Series in Computational Mathematics, vol. 31. Springer, Berlin (2006)
32. Hutchings, J.A.: Adaptive phenotypic plasticity in brook trout, Salvelinus fontinalis, life histories. Ecoscience **3**, 25–32 (1996)
33. Kaczensky, P., Chapron, G., von Arx, M., Huber, D., Andrn, H., Linnell, J.: Status, Management and Distribution of Large Carnivores 'Bear, Lynx, Wolf and Wolverine' in Europe, European Commission (2013, March)
34. Kot, M.: Elements of Mathematical Ecology. Cambridge University Press, Cambridge (2001)

35. Lacitignola, D., Diele, F., Marangi, C.: Dynamical scenarios from a two-patch predator prey system with human control - implications for the conservation of the wolf in the Alta Murgia National Park. Ecol. Model. **316**, 28–40 (2015)
36. Lacitignola, D., Diele, F., Marangi, C., Provenzale, A.: On the dynamics of a generalized predator-prey system with Z-type control. Math. Biosci. **280**, 10–23 (2016)
37. Lenhart, S., Workman, J.T.: Optimal Control Applied to Biological Models. Chapman & Hall/CRC, London (2007)
38. Leonard, D., Van Long, N.: Optimal Control Theory and Static Optimization in Economics. Cambridge University Press, Cambridge (1992)
39. Li, X., Yong, J.: Optimal Control Theory for Infinite Dimensional Systems. Birkhauser, Boston (2001)
40. Lions, J.L.: Optimal Control of Systems Governed by Partial Equations. Springer, Berlin (1970)
41. Magnea, U., Sciascia, R., Paparella, F., Tiberti, R., Provenzale, A.: A model for high-altitude alpine lake ecosystems and the effect of introduced fish. Ecol. Model. **251**, 211–220 (2013)
42. Marescot, L., Chapron, G., Chades, I., Fackler, P.L., Duchamp, C., Marboutin, E., Gimenez, O.: Complex decisions made simple: a primer on stochastic dynamic programming. Methods Ecol. Evol. **4**(9), 827–884 (2013)
43. Marinoschi, G., Martiradonna, A.: Fish populations dynamics with nonlinear stock-recruitment renewal conditions. Appl. Math. Comput. **277**, 101–110 (2016)
44. Martiradonna, A., Diele, F., Marangi, C.: Analysis of state-control optimality system for invasive species management. In: Analysis, Probability, Applications, and Computation, pp. 3–13. Birkhäuser, Cham (2019)
45. Materassi, M., Innocenti, G., Berzi, D., Focardi, S.: Kleptoparasitism and complexity in a multi-trophic web. Ecol. Complex. **29**, 49–60 (2017)
46. May, R.M.: Stability and Complexity in Model Ecosystems. Princeton University Press, Princeton (1973)
47. McGeoch, M.A., Butchart, S.H.M., Spear, D., Marais, E., Kleynhans, E.J., Symes, A., Chanson, J., Hoffmann, M.: Global indicators of biological invasion: species numbers, biodiversity impact and policy responses. Divers. Distrib. **16**(1), 95–108 (2010)
48. Monitoraggio dei carnivori nel parco nazionale dell'Alta Murgia primi risultati. Parco nazionale dell'Alta Murgia, Servizio Tecnico (2010)
49. Murray, J.D.: Mathematical Biology I. An Introduction. Springer, Berlin (2002)
50. Odom, D.I.S., Cacho, O.J., Sinden, J.A., Griffith, G.R.: Policies for the management of weeds in natural ecosystems: the case of scotch broom (Cytisus scoparius, L.) in an Australian national park. Ecol. Econ. **44**, 119–135 (2003)
51. Pontryagin, L.S., Boltyanskii, V.G., Gamkrelize, R.V., Mishchenko, E.F.: The Mathematical Theory of Optimal Processes. Wiley, New York (1962)
52. Sanz-Serna, J.M.: Symplectic Runge-Kutta schemes for adjoint equations, automatic differentiation, optimal control, and more. Siam Rev. **58**, 3–33 (2016)
53. Simberloff, D.: We can eliminate invasions or live with them. Successful management projects. Biol. Invasions **11**(1), 149–157 (2009)
54. Spring, D., Cacho, O.J.: Estimating eradication probabilities and trade-offs for decision analysis in invasive species eradication programs. Biol. Invasions **17**(1), 191–204 (2014)
55. Strogatz, S.H.: Nonlinear Dynamics and Chaos: With Applications to Physics, Biology, Chemistry, and Engineering. Studies in Nonlinearity, 1st edn. Westview Press, Boulder (2001)
56. Tiberti, R., Acerbi, E., Iacobuzio, R.: Preliminary studies on fish capture techniques in Gran Paradiso alpine lakes: towards an eradication plan. J. Mt. Ecol. **9**, 61–74 (2013)
57. Turchin, P.: Complex Population Dynamics: A Theoretical/Empirical Synthesis. Monograph in Population Biology. Princeton University Press, Princeton (2003)
58. Zhang, Y., Li, Z.: Zhang neural network for online solution of time-varying convex quadratic program subject to time-varying linear-equality constraints. Phys. Lett. A **373**(18), 1639–1643 (2009)

59. Zhang, Y., Yi, C.: Zhang Neural Networks and Neural-Dynamic Method. Nova Science Publishers, New York (2011)
60. Zhang, Y., Yan, X., Liao, B., Zhang, Y., Ding, Y.: Z-type control of populations for Lotka-Volterra model with exponential convergence. Math. Biosci. **27**, 15–23 (2016)

Printed in the United States
by Baker & Taylor Publisher Services